Toxic Effects of Nanomaterials

Edited By

Haseeb Ahmad Khan

King Saud University
Saudi Arabia

Ibrahim Abdulwahid Arif

King Saud University
Saudi Arabia

CONTENTS

About the Editors

Dr. Haseeb Ahmad Khan is a Chair Professor at Prince Sultan Research Chair for Environment and Wildlife, College of Science, King Saud University, Riyadh, Saudi Arabia. Before joining this position, he served as a Senior Scientist at Armed Forces Hospital, Riyadh, Saudi Arabia and then as an Associate Professor of Biochemistry at King Saud University, Riyadh. He obtained his PhD from Aligarh Muslim University, Aligarh, India and received scientific trainings at USA and UK. He is a Fellow of the Royal Society of Chemistry, UK and a Member of American Chemical Society, USA and Royal Australian Chemical Institute, Australia. He is an editorial board member and reviewer for several international journals. He has published more than 100 research papers, authored six book chapters and filed two patents. He has also developed seven software tools for biomedical applications. His multidisciplinary research interests include toxicology, analytical biochemistry, drug interactions, molecular diversity and bioinformatics.

Prof. Ibrahim Abdulwahid Arif is a Professor of Biology at the Department of Botany and Microbiology, College of Science, King Saud University, Riyadh, Saudi Arabia. Recently, he has been appointed as a Consultant at the Ministry of Higher Education. He is also holding the authoritative positions of the Supervisor of Prince Sultan Research Chair for Environment and Wildlife and the President of Saudi Biological Society. He completed his MSc from Colorado State University and PhD from the University of Utah and started his professional career as a faculty member at King Saud University. During his tenure, he has also served as a Deputy Dean of the College of Science as well as the Chairman of various administrative committees. He has published about 50 research papers in international journals, authored 10 scientific books and filed two patents. His achievements in environmental issues are noteworthy due to his significant contribution in environmental conservation by studying the multi-factorial biological interactions with the environment.

FOREWORD

Biomedical, technological, and consumer product oriented applications of engineered nanomaterials is an expanding global phenomenon that is outpacing our scientific and quantitative understanding of potential toxicological consequences from nanomaterial exposure. This has hastened on an international scale the intensity of efforts to investigate the environmental and human health and safety risk from nanomaterial exposure. In 2001 the United States established the National Nanotechnology Initiative (NNI) program to coordinate National nanotechnology research. Since 2008 a key NNI focus has been on developing a National strategy for assessing nanotechnology-related environmental, health, and safety research. In March 2009 the European Union (EU) was the first in the world to establish strict regulatory guidelines imposing stringent workplace practices and requiring that nanomaterials be proven safe before used in products to prevent exposure to potentially toxic materials.

Considerable challenges exist however, in the endeavor to quantify the safety of nanomaterials given their diversity in terms of size, shape, charge, composition, solubility, and the wide range of processes used to synthesize and incorporate them into products. There is an added challenge that nanomaterials maybe transformed upon contact with biological systems and in the environment. Overcoming these challenges is encumbered by the lack of sensitive analytical techniques to detect and quantify nanomaterials in biological matrices, the environment, and in the workplace. This has spurred efforts to develop standardized nanomaterials for use in toxicological testing as well as standardized methods to characterize their physicochemical properties. Again the EU is leading this charge as in February 2011 the European Commission's Joint Research Centre launched the first European repository of nanomaterials comprising of 25 nanomaterial standards. These can be used to investigate important questions as to which physicochemical properties most affect nanomaterial interactions with epithelial tissues, cytotoxicity, and transport through biological and environmental systems.

Comprehensive documentation of results as presented in this book, "**Toxic Effects of Nanomaterials**", is one important mechanism to communicate the status of this emerging field of Nanotoxicology for which few previous examples exist. A key feature of this book is inclusion of studies investigating potential toxicity of nanomaterial exposure to plants and aquatic life in the environment for which far less is known as the research community has focused mostly on assessing human health concerns particularly from skin and respiratory exposure. In summary, this timely book presents a state of the art view of all aspects of this complex Nanotoxicology field and is a contribution that will serve as a foundational guide in this field and the truth about the safety of nanomaterials evolves.

Lisa A. Delouise
Departments of Dermatology and Biomedical Engineering
University of Rochester Medical Center, USA

PREFACE

This decade has seen revolutionary developments in the field of nanotechnology with newer and diverse applications of nanoparticles appearing everyday. Novel nanomaterials are emerging with different characteristics and compositions for specific applications such as cosmetics, drug delivery, imaging, electronic *etc*. However, little attention is being paid to understand, assess and manage the environmental impact and adverse effects of nanoparticles. Currently the information about the toxicity of nanoparticles and their environmental fate in air, water, soil and tissues is severely lacking. Inhalation, ingestion and dermal penetration are the potential exposure routes for nanoparticles whereas particle size, shape, surface area and surface chemistry collectively define the toxicity of nanoparticles. Several studies have shown excessive generation of reactive oxygen species as well as transient or persistent inflammation following exposure to various classes of nanoparticles. Increased production and intentional (sunscreens, drug-delivery, *etc.*) or unintentional (environmental, occupational, *etc.*) exposure to nanoparticles increases the possibilities of adverse health effects. Thus, the novel nanomaterials need to be biologically characterized for their health hazards to ensure risk-free and sustainable implementation of nanotechnology. Currently there are only a few books available in this specific area to cover toxicological aspects of nanoparticles. A reasonably priced, comprehensive book on nanotoxicology was therefore badly needed by the nanocommunity to clearly understand the subject and we tried fulfill their demand.

The present book "Toxic Effects of Nanomaterials", comprised of 8 chapters with 77 illustrations (60 figures and 17 tables), provides an authoritative work of international experts in the field of nanotoxicology. The most important feature of the book is a broad coverage of phytotoxicity of nanoparticles, which is largely neglected in many texts. The first two chapters of this book deal with the toxicity of nanoparticles in plants. The third, fourth and fifth chapters discuss the toxicities of iron oxide, titanium oxide and silicon oxide nanoparticles. The sixth chapter provides a comprehensive review of methodologies used in nanotoxicology. The last two chapters highlight the risks associated with the use of nanoparticles and the environmental impact of nanomaterials. Such a broad coverage of nanotoxicology renders this book highly beneficial to the scientists from multidisciplinary areas including nanotechnologists, toxicologists, pharmacologists, environmental chemists and biomedical scientists. This book would equally be useful for the individuals advocating for sustainable use of nanotechnology. We are thankful to all the eminent scientists who have contributed their chapters to this book. The publishing platform provided by the Bentham Science Publishers is gratefully acknowledged.

Haseeb Ahmad Khan
Ibrahim Abdulwahid Arif

List of Contributors

ABDULLAH SALEH ALHOMIDA

Department of Biochemistry, College of Science
King Saud University, Riyadh
SAUDI ARABIA
E-mail: alhomida@ksu.edu.sa

ADNAN HAIDER

Department of Chemistry
Kohat University of Science and Technology, Kohat
PAKISTAN
E-mail: adnan_afridi9294@yahoo.com

ALFREDO MIRANDA GOES

Department of Biochemistry and Immunology
ICB, UFMG, Campus-Pampulha
Belo Horizonte, Minas Gerais
BRAZIL
E-mail: goes@mono.icb.ufmg.br

AMITAVA MUKHERJEE

Nanobiomedicine Lab
School of Biosciences and Technology
VIT University, Vellore
INDIA
E-mail: amitav@vit.ac.in

ANGELA LEAO ANDRADE

Department of Chemistry, ICEB
Federal University of Ouro Preto
Ouro Preto, Minas Gerais
BRAZIL
E-mail: angelala01@hotmail.com

ANNA SPERANZA

Dipartimento di Biologia ES
Università di Bologna, Bologna
ITALY
E-mail: anna.speranza@unibo.it

FATIMA KHANAM

Department of Chemistry, College of Science
King Saud University, Riyadh
SAUDI ARABIA
E-mail: fatimakhanam4@gmail.com

HASEEB AHMAD KHAN

Prince Sultan Research Chair for Environment and Wildlife
King Saud University, Riyadh
SAUDI ARABIA
E-mail: khan_haseeb@yahoo.com

IBRAHIM ABDULWAHID ARIF

Prince Sultan Research Chair for Environment and Wildlife
King Saud University, Riyadh
SAUDI ARABIA
E-mail: iaarif@hotmail.com

JOSE DOMINGOS FABRIA

Department of Chemistry, ICET
Federal University of Jequitinhonha and Mucuri Valleys
Diamantina, Minas Gerais
BRAZIL
E-mail: jdfabris@ufmg.br

KERSTIN LEOPOLD

Fachgruppe für Analytische Chemie
Technische Universität München
Lichtenbergstrasse, Garching
GERMANY
E-mail: kerstin.leopold@ch.tum.de

LAURA MANODORI

Veneto Nanotech, via San Crispino 106, Padua
ITALY
E-mail: laura.manodori@venetonanotech.it

LISA BREGOLI

Veneto Nanotech, via San Crispino 106, Padua
ITALY
E-mail: lisa.bregoli@ecsin.eu

LUCAS REIJNDERS

IBED, University of Amsterdam
Nieuwe Achtergracht, Amsterdam
THE NETHERLANDS
E-mail: l.reijnders@uva.nl

MAMTA KUMARI

Nanobiomedicine Lab
School of Biosciences and Technology
VIT University, Vellore
INDIA
E-mail: mamta_kiku@yahoo.co.in

MOHAMMAD ABDUL BAKIR

Prince Sultan Research Chair for Environment and Wildlife
King Saud University, Riyadh
SAUDI ARABIA
E-mail: mabakir@yahoo.com

MORTEZA MAHMOUDI

National Cell Bank, Pasteur Institute of Iran
Institute for Nanoscience and Nanotechnology
Sharif University of Technology, Tehran
IRAN
E-mail: mahmoudi@biospion.com

N. CHANDRASEKARAN

Nanobiomedicine Lab
School of Biosciences and Technology
VIT University, Vellore
INDIA
E-mail: nchandra40@hotmail.com

NAUSHEEN BUKHARI

Department of Chemistry, College of Science
King Saud University, Riyadh
SAUDI ARABIA
E-mail: nenosahil@hotmail.com

ROSANA ZACARIAS DOMINGUES

Department of Chemistry, ICEx, UFMG
Campus-Pampulha, Belo Horizonte, Minas Gerais
BRAZIL
E-mail: rosanazd@qui.ufmg.br

SAJJAD HAIDER

Department Chemical Engineering, College of Engineering
King Saud University, Riyadh
SAUDI ARABIA
E-mail: shaider@ksu.edu.sa

SALMAN AL ROKAYAN

King Abdullah Institute for Nanotechnology
King Saud University, Riyadh
SAUDI ARABIA
E-mail: dr.salman@alrokayan.com

SOPHIE LAURENT

Department of General, Organic and Biomedical Chemistry
NMR and Molecular Imaging Laboratory, University of Mons
BELGIUM
E-mail: sophie.laurent@umons.ac.be

STEFANO POZZI-MUCELLI

Veneto Nanotech, via San Crispino 106, Padua
ITALY
E-mail: stefano.pozzimucelli@venetonanotech.it

VINITA ERNEST

Nanobiomedicine Lab
School of Biosciences and Technology
VIT University, Vellore
INDIA
E-mail: rach.jun@gmail.com

W. SHANE JOURNEAY
Nanotechnology Toxicology Consulting and Training, Inc., Nova Scotia
Faculty of Medicine, Dalhousie Medical School
Dalhousie University, Halifax
CANADA
E-mail: journeay@dal.ca

Toxic Effects of Nanomaterials

2

CHAPTER 1

Nanoparticle-Induced Toxicity: Focus on Plants

Less is more
(Robert Browning, 1812-1889)

Anna Speranza[1*] and Kerstin Leopold[2]

[1]*Dipartimento di Biologia ES, Università di Bologna, Bologna, Italy and* [2]*Fachgruppe für Analytische Chemie, Technische Universität München, Lichtenbergstraße, Garching, Germany*

Abstract: Nanoparticle technology offers a large array of applications also in plants, for either plant biology research or agricultural practice. Indeed, plants are at the base of any ecological web, in both natural and artificial ecosystems; nanoparticles of various origins, natural as well as anthropogenic or engineered, are being increasingly released into the environment. Therefore, assessment of possible risks is urgent before nanoparticles become more and more ubiquitous in every aspects of life. The present chapter critically reviews recent knowledge on phytotoxicity of nanoparticles, considering both lower and higher plants.

Keywords: Natural NPs, Anthropogenic NPs, Engineered NPs, Carbon NPs, Metal NPs, Metal oxide NPs, Oxidative stress, Uptake by plants, Translocation in plants, Phytotoxicity, Microalgae, Seed germination, Pollen germination.

INTRODUCTION

Nanoparticles (NPs) are defined as particulate matter with at least one dimension that is less than 100 nm [1, 2]. Thereby, the particle consists of either atomic or molecular aggregates ranging in their scale between the atomic/molecular level and larger scale bulk material. Since the beginning of the earth there have always been natural NPs in the atmosphere, hydrosphere, and soil, such as salt aerosols from sea water spills, hydro colloidal clays, humic matter, volcanic dust, lunar dust, mineral composites, *etc.* Furthermore, with the industrialization anthropogenic NPs caused by coal combustion, traffic, and industrial processes, were unintended formed and emitted into the environment. The sum of all airborne anthropogenic particulate matter is often referred to as "fine dusts" or "air pollution particles" being a mixture of organic and inorganic particles with a size range from the nanometer scale up to 10μm [3, 4]. However, the majority of anthropogenic NPs are carbon-based and result from incomplete incineration processes, such as soot from coal combustion, diesel exhausts, fly ash, tar leachates, *etc.* [5]. Moreover, carbon black is also emitted by *e.g.,* abrasion from tires [6]. Inorganic anthropogenic NPs are often formed by corrosion and abrasion processes. Metal oxide NPs, such as TiO_2 for instance, can be found in waters from exterior facades and roof run-offs [7]. Metal and metal oxide NPs can be emitted from bearings and brakes [8]. Furthermore, automotive exhaust catalytic converters release platinum group metal NPs (mainly platinum, Pt; palladium, Pd; and rhodium; Rh) which are applied as main catalytic active elements for the conversion of carbon monoxide, nitrogen oxides, and hydrocarbons into non-hazardous gases. About 90% of these emissions are particulate matter varying in size from >10.2μm to <5nm. The coarse particles consist of alumina and silica particles that carry dispersed platinum group metal (PGM) NPs [9-11]. The fine particles, *i.e.,* the emitted NPs, are mainly elemental metal and can make up to a third of the total PGM emission [12-14]. Initially, PGM contamination occurs in airborne particulate matter, roadside dust, soil, sludge, and waters [15]. Finally, these metals can also be taken up by biota and thus entering the life cycle [16]. Among the platinum group elements Pd is known to be the most reactive element with the highest mobility in the environment and elevated concentrations of Pd have not only been found in road dusts [17]

Address correspondence to Anna Speranza: Associate Professor, Dipartimento di Biologia ES, Università di Bologna, Bologna, Italy; Tel: +39(0)51-2091314; Fax +39(0)51-242576; E-mail: anna.speranza@unibo.it and top soils, but also in deeper soil layers [18]. Therefore, the uptake and effects of these particles by plants and pollen are of interest and discussed in detail in section 6.

Within the last decades a third group of NPs, engineered NPs, were produced because of their unique properties that differ significantly from those of the corresponding bulk materials (see section 2). A broad variety of technological fields benefits from these materials, *e.g.,* human health (medical imaging, diagnosis, drug delivery, hygiene), energy (improved efficiency, catalysis, hydrogen storage), environmental technology (water filtration, remediation), and agriculture (secure packaging, increased crop yields). Furthermore, NPs are applied in diverse consumer products, like tires, stain-resistant and antimicrobial clothing, sporting goods, sunscreens, detergents, cosmetics, surfactants, dyes, pigments, and electronics. Consequently, nanotechnology is an emerging field with a total global investment of around $10 billion in 2005 [19] and it is estimated that the annual value for all nanotechnology-related products will be $1 trillion by 2011-2015 [20]. While not comprehensive, currently about 1, 000 NPs-containing products are registered in the "Project on Emerging Nanotechnologies" from the Woodrow Wilson Centres (USA) [21], where data based on voluntary information given by the manufacturers is collected. An estimation for the production of nanomaterials in 2004 was 2000 t and for the years 2011-2020 an increase to about 58, 000 t is expected [22]. The present multitude of different engineered NPs can be separated into 4 main groups:

1. Carbon NPs: Nanoparticles based on carbon, like fullerene and carbon nanotubes.

2. Metal NPs, metal oxide and sulfide NPs, such as nanosilver, nanogold, quantum dots (QDs), nanoaluminum, and nano metal oxides like TiO_2, ZnO, Al_2O_3.

3. Dendrimers: Nano-sized polymers of repeatedly branched molecules.

4. Composites: A combination of nanoparticles with other nanoparticles or with bulk materials.

The most common engineered NP is nanosilver which is mainly applied for antimicrobial purposes in *e.g.,* wound dressing, detergents, and functional wear and can be found in approximately 20% of all NPs-containing products [23]. Other widely used NPs are zerovalent iron, applied in groundwater remediation as an effective tool for degrading a wide variety of common contaminants [24], ZnO NPs in sunscreens [25], and TiO_2 NPs that are used as photocatalysts (in water and wastewater treatment), pigments and cosmetic additives (as UV quencher and whitening agent), in sunscreens and food products [26]. Furthermore, QDs are fluorescent nanocrystals (10 to 25 nm diameter): they have a protective organic coating and an inner core containing cadmium and selenium. If intact, the QD coating protects against toxic core metal; however, the surface coating is subjected to photolysis or oxidation, resulting in toxic metal release from the core [27].

Emission of engineered NPs into the environment can occur during their production, application, and disposal. For instance, studies on the leaching of silver NPs from sport socks that contained up to 1360 µg Ag g^{-1} sock showed amounts of 650 µg of silver in 500 mL washwater [28]. Hence, a considerable amount of nanosilver can be found in domestic sewage and enters wastewater treatment plants where most of it is removed from the water and transferred to the sewage sludge. Finally, disposal of the sludge as agricultural fertilizer brings nanosilver back to the environment where it has toxic effects on soil microbes [29]. Furthermore, high amounts of Ag NPs in domestic effluents can hamper nitrogen degradation in wastewater treatment plants by affecting useful bacteria and thus interfering effective biological treatment of wastewaters. For the European Union an estimation of cumulative aquatic exposure and risk due to silver emission including new nanosilver products has been published for the year 2010 [30]. For example, for the river Rhine it is estimated that in 2010 about 15% of the silver will be silver NPs originating from nanoproducts.

However, long-term distribution of any NPs will always lead to immission in the hydrosphere and soils. Hence, biota will be exposed to increasing amounts of engineered NPs within the next decades (Fig. **1**).

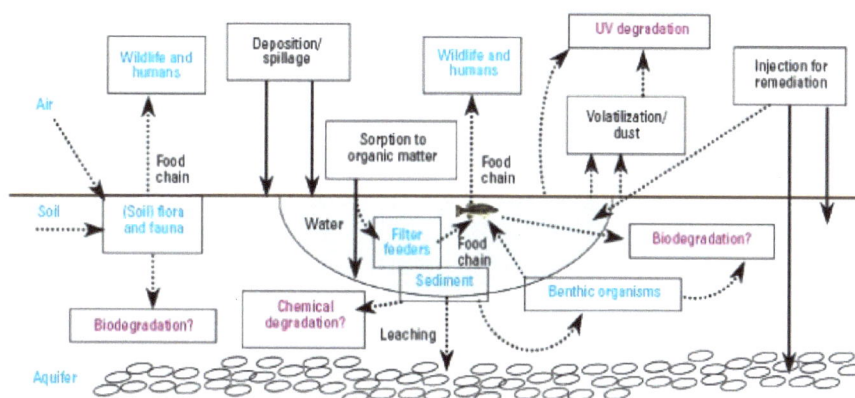

Figure 1: Possible pathways of NPs in the environment. Solid lines represent routes that have achieved experimental evidence or that are currently in use (remediation). Magenta and blue colour refers to possible degradation routes and sinks, respectively. (Reproduced from [31] with permission from Environmental Health Perspectives)

Studying the potential adverse effects of NPs on human health, biota, and ecology has become a top issue in the scientific community within the last decade [32]. In 2006 the first international conference on "Environmental Effects of Nanoparticles and Nanomaterials" was held in London and conferences took place yearly since then. Moreover, the discussion on potential risks arising from nanotechnology has become a top concern in governments and public all over the world. The United States Environmental Protection Agency (US EPA), the European Community, as well as many other authorities recommend further research for comprehensive risk assessment of nanomaterials [33, 34]. For this purpose, exposure scenarios and emission pathways have to be investigated along with toxicological effect studies. Thereby, the investigation of uptake and effects of NPs on plants is a key issue in assessing nanotechnology's impact on our environment. The present review discusses scale-dependent chemical and physical properties and explains nanotoxicity at the biochemical level. Moreover, it gives a comprehensive summary on up-to-date knowledge on NPs-induced toxicity on plants.

SCALE-DEPENDENT PROPERTIES

The main characteristic of NPs is their extremely small size and with it their great surface area. There are several reasons why the physical and chemical properties of a material can change with decreasing size. As particles decrease in size, the proportion of atoms found at the surface to those that comprise the interior increases tremendously (*cf.* Fig. **2**). In some materials almost every atom of the nanoparticle is exposed on the surface (*e.g.,* single-wall carbon nanotubes and fullerenes) [35]. Hence, quantum effects are more important in determining the properties and characteristics of NPs. Furthermore, the high surface free energy and the high radius of curvature of a NP may alter its thermodynamic properties. Consequently, NPs can show appreciable different chemical and physical properties compared to the same material at larger scale [36]. Some specific chemico-physical properties of NPs that may have an impact on their environmental and toxicological effects are discussed briefly in the following paragraphs.

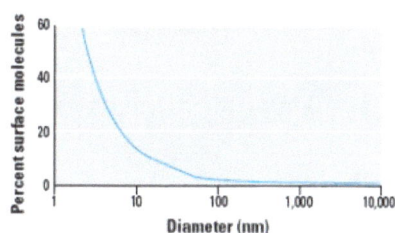

Figure 2: Surface molecules as a function of particle size. (Reproduced from [31] with permission from Environmental Health Perspectives)

Solubility and Mobility

Solubility and mobility of NPs might be the most important issues when considering their impact on the environment in the first place. Because of their small size nanoparticles follow Brownian motion in solutions and therefore can form dispersions, *i.e.,* colloidal solutions. Hence, their "actual solubility" is by far greater than the chemical solubility of dissolved ions or molecules of the same composition. At the same time, the mobility of NPs in a specific media (diffusion) increases with decreasing size of the particle. In fact, the diffusion rate of a particle with 50 nm diameter is almost one order of magnitude lower than that of a particle with a diameter of 0.5 nm [36]. However, when there is no barrier for a close approach of the particles, such as surface charge or steric stabilisation, the high surface energy of NPs leads to agglomeration of the particles when colliding. In the environment NPs interact with natural air and water compounds, such as gaseous compounds or dissolved ions, as well as with surfaces, like soils and natural NPs. Hence, adsorption of NPs to soils or sediments may occur as well as coating of the NPs surface with organic and inorganic matter. These very complex processes occurring in the environment are at presence poorly understood under real conditions and are subject to detailed investigations [37]. The knowledge on the behaviour of natural aquatic colloid and atmospheric ultrafine particles may help to assume possible processes, but further direct evidence is required.

Finally, the role of NPs as possible transport vehicles for different substances in the environment as well as in living organisms has to be mentioned in respect to mobility of NPs. For natural hydro colloids the phenomena of transportation of abundant pollutants has been widely described [38]. Sparingly soluble pollutants that are normally retained by adsorption onto soil or sediment surfaces regain high mobility in the presence of hydro colloids. They "travel" adsorbed onto the particles surface and thus cover long distances in the environment. Anthropogenic and engineered NPs can presumably show similar behaviour and thereby increase the mobility and possibly the bioavailability of pollutants. On the other hand, this effect has been used to deliver *e.g.,* drugs or deoxyribonucleic acid (DNA) to particular positions in living organisms, *i.e.,* in animals and plants [39, 40].

Catalytic Activity and Reactivity

Due to the high surface area to volume ratio of NPs the number of atoms available for chemical reactions is higher than for the same material at larger scale. This fact increases the chemical reactivity and catalytic activity of nanomaterials. In some cases, materials that are "non-active" at large scale, *e.g.,* gold [41], show significant catalytic activity when applied as NPs [42]. Enhanced chemical reactivity is a benefit for industrial processes, but at the same time might cause adverse effects when brought into a (natural) biochemical environment.

Electronic and Optical Properties

The electronic system of an atom is distinctly different from that of a solid material. With an increasing number of atoms forming a crystal lattice the energetic levels of the orbitals change. In quantum dots, for instance, the intrinsic band gap of the semiconductor is affected in a way that photoluminescence occurs [43]. In the same way magnetic properties of NPs can differ from those of bulk materials and magnetism is suspected to cause genetic changes in the biochemical environment.

Thus, all these properties can lead to significantly different toxicity in comparison to the same material at larger scale. At the same time NPs approach the length scale at which increased uptake and interaction with biological tissues are likely [44].

MECHANISMS OF NANOTOXICITY AT THE CELL LEVEL

Impact of NPs to cells elicits a number of responses: so far, these have been described referring almost exclusively to animal cells, including humans. Therefore, papers quoted in this section necessarily disregarded plants. However, it is tempting to hypothesize that at least a part of intimate mechanisms of nanotoxicity operating in animal cells may be quite close to those of plant cells. The main structural elements and their functions are very similar, in fact, in the majority of eukaryotic cells [45].

Basically, studying *in vitro* interactions with isolated, functional biomolecules is promising as a preliminary understanding of toxicity mechanisms. In this view, nano-TiO_2 was selected as representative of NPs and lysozyme as a representative of enzymes. Effects on both spatial structure and bacteriolytic activity of the enzyme were revealed. It was concluded that nano-TiO_2, whose genotoxic and cytotoxic effect in cultured human cells are known, might have impact on natural immunity *in vivo* [46].

At the cell level, oxidative stress in relation to NP toxicity has been largely documented. Oxidative stress consists of imbalance between production of reactive oxygen species (ROS) or reactive nitrogen species (RNS), which are harmful for biomolecules, and activation of the antioxidant defence machinery (*e.g.,* radical scavengers, antioxidant enzymes) aimed to counteract them. Several studies document a strict association between exposure to NPs and induction of oxidative stress [47-51]. For instance, ROS generation has been proposed to explain toxicity of inhaled NPs [52]. To the best of our knowledge, only one report is available on induction of oxidative stress in plant cells by NPs: activation of antioxidant genes was in fact shown in a green unicellular alga spiked with TiO_2 and QDs [53].

Different levels of oxidative damage produced by different types of NPs in animal systems were exhaustively reviewed [54]. These included well-characterized preparations of unintended formed anthropogenic NPs (air pollution particles, wood smoke particles, and diesel exhaust particles); carbon NPs (such as C_{60} fullerenes, *i.e.,* 60 carbon atoms arranged in a spherical structure like a soccer ball, and carbon nanotubes); metal and metal oxide NPs (silica, titanium dioxide). First of all, it has to be preliminary remarked that many NPs are *per se* capable of producing toxic ROS, and their ability to generate ROS in cell-free assays has been described; this, obviously, may have detrimental effects to ecosystem health [54, 55]. Various factors influence the amount of ROS produced. Air pollution particles can generate high levels of ROS in the presence as well as in the absence of H_2O_2 [56-58]. C_{60} fullerenes have only a small ability to generate ROS when suspended directly in aqueous solutions, while the ability increases with previous suspension in tetrahydrofuran [59, 60]. As for carbon nanotubes, single-walled carbon nanotubes (SWCNTs) produce concentration-dependent ROS generation; not only multi-walled carbon nanotubes (MWCNTs) do not generate ROS but, evenly, they exhibit radical scavenging activity [59, 61].

The majority of studies on cultured cells indicate that exposure to air pollution particles, diesel exhaust particles, and carbon nanotubes increases intracellular ROS production [59, 62-69]; only a few reports deny it [70, 71]. As for metal NPs, it was shown that increase in ROS level was an important mediator of cytotoxicity caused by Ag-nanoparticles in a human acute monocytic leukemia cell line (THP-1), where apoptosis and necrosis were also induced [72]. Moreover, gold NPs were able to induce necrosis by oxidative stress and mitochondrial damage in He-La cells [73].

A further index of oxidative stress is cell depletion of defence compounds designed to cope with the stress; for example, the ubiquitous glutathione is an endogenous, reducing agent which is depleted by ROS damage. A concentration-dependent decrease in the ratio of reduced glutathione (GSH) over oxidized glutathione disulphide (GSSG) has been in fact demonstrated in several cell lines exposed to diesel exhaust particles [54, 74, 75]. Studies on air pollution particles showed that the GSH/GSSG ratio decreased in cells exposed to fine and ultrafine size fraction, but not in the case of coarse particles [76]. Exposure to SWCNTs containing about 30% iron also depleted the endogenous GSH [77]. Furthermore, consistent with the fact that increased production of ROS may be associated with mitochondrial dysfunction [78], ultrafine air pollution particles were found to induce mitochondrial damage [76]. Also diesel exhaust particles, various particulates derived from ambient street air, and wood burning affected mitochondria [79]. By contrast, C_{60} fullerenes did not interfere with mitochondrial activity, while inducing lipid peroxidation, that is, oxidative damage to lipids, which are essential constituents of cell membranes [80]. Lipid peroxidation is a mechanism of cell injury common to both animal and plants. Many reports document lipid peroxidation produced by exposure of cells to silica, air pollution particles, wood smoke particles, C_{60} fullerenes, and SWCNTs [54]. Adverse effects induced by C_{60} fullerenes to different types of human cells *via* lipid peroxidation were documented [80]. Moreover, a number of studies described genotoxicity of NPs; for instance, the highest ability to induce DNA damages is exhibited by diesel exhaust particles and other air pollution particles, while C_{60} fullerenes exhibit the lowest ability [54]. Besides oxidative mechanism, metal

NPs were found to induce neurotoxicity in PC12 neuronal cell line; they induced dopamine depletion also through enzymatic alteration, in particular, by down-regulation of several genes associated with the dopaminergic system [81].

As for biochemical alterations induced by traffic related emissions, it was shown that exposure to nanoparticle-rich diesel exhaust (NR-DE) resulting from diesel engine-powered cars and trucks caused neurotoxicity in the mouse olfactory bulb by significantly increasing extracellular glutamate level while, concomitantly, gene expression of memory-related functions was upregulated [82]. It is also worth noticing to mention the area of reproductive toxicity: it was already known that traffic pollutants affect the fertility of men employed at the motorway tollgates [83]. As for the particulate traffic emissions, exposure to NR-DE increased serum and testis testosterone levels in rats [84]. The mechanism of such an alteration was further investigated; thus, it resulted that disruption of testosterone biosynthesis due to NR-DE exposure involved elevation of steroidogenic acute regulatory protein and cytochrome P450 side chain cleavage messenger ribonucleic acid (mRNA) and their protein expression in the testis [85].

DEAD OR ALIVE: PLANTS CAN PRODUCE NPS THEMSELVES

It is perhaps unexpected that plant can function as NP factory. Recent reviews exhaustively illustrate production of precious metal NPs by dead plant biomasses (algae and higher plants), tissue extracts from several plant species, isolated plant cells or entire plants. Many different functional groups within the biological material (from sugars, carboxyl acids of cell walls, carbonyl groups, amine groups) may be involved in the process [86-88]. For instance, biomass from alfalfa (*Medicago sativa*) and oat (*Avena sativa*) is able to reduce Au(III) to Au(0) in solution forming Au NPs [89, 90]. The growth of Au NPs occurring inside live plants of alfalfa has been described [91]; the nucleation of the particles seemed to be located in preferential zones. Moreover, an active transport of Au(0) from roots to shoots was suggested to occur throughout the plant. TEM image of the alfalfa shoots confirmed the existence of Au NPs aggregated together as small black dots. This has important implications and possible applications; for example, it provides an alternative, inexpensive method for the synthesis of gold NPs. Furthermore, it indicates that a live plant, alfalfa, can perform phytomining, that is, extraction of valuable, naturally occurring elements of soil [91]. Silver hyperaccumulation in plants was for the first time described in *Medicago sativa* and *Brassica juncea*, which were able to accumulate the metal (up to 13.6 and 12.4%, respectively) when exposed to an aqueous substrate containing 1, 000 ppm $AgNO_3$ for 24 h; in both species, silver was stored as NPs of about 50 nm [92]. Moreover, Pd NPs ranging from 3.2 to 6.0 nm were produced by leaf broth from *Cinnamomum camphora* [93]. Evenly, plants can form mixed-metal (Au, Ag, and Cu) NPs having catalytic potential, as it was shown in Indian mustard [94]. It seems, finally, that the characteristics of biologically and chemically synthesized precious metal NPs are essentially the same [88].

NPS AS SMART DELIVERY SYSTEMS IN PLANTS

Several of the studies which are mentioned in next sections address the question of NP uptake into entire plants or plant cells. In general, uptake is at the basis of very interesting application of NPs, that is, the use as 'smart' delivery systems in living beings. In the exciting, rapidly developing field of nanomedicine, NPs provide much earlier and more accurate diagnoses, far less invasive procedures, and have the ability to control topography and chemistry at the nanoscale level, therefore making sure that the corrected dose of a drug reaches the intended tissue or organ in a patient [95, 96]. The basic principle of curing with the aid of NPs can be applied also to plant diseases, for instance to prevent or tackle infections. NPs containing fungicides were successfully applied to southern yellow pine and birch wood. The method had unexpected high efficiency, probably because the NPs served as reservoirs, allowing controlled release of the biocides and therefore increasing wood preservation [97-99].

Insofar, studies which were aimed to specifically investigate uptake of NPs by plants and follow their fate throughout plant tissues are not much numerous. Concern of some authors [100-102] as for the possibility that NPs can overstep the barrier of cell wall, which is a complex, macromolecular network outside plant cells, seems not to be substantial. Transport and deposition of NPs inside pumpkin plants was visualized and tracked [103]. Suspensions of carbon-coated iron NPs were injected into the leaf petiole; the NPs were found to have been translocated to other plant districts through the vascular system, which has vessels much large in diameter, and internalized into plant tissues. Due to the magnetic character of the NPs, it was possible to control their pathways throughout the plant by applying a magnetic field [103]. This technique offers a huge range of possibilities, in the view of charging the NPs with different substances to be introduced within plants and then to be directed and concentrated to desired areas. This is particularly interesting in the field of plant research; as for crops, it is surely impractical to place magnets to individual plants such as cereals, while specific treatments to trees are well reasonable [103]. No damage to the treated pumpkins plants was observed, though local damage to some cells could not be excluded; however, the possibility that more intense treatments could be phytotoxic might not be definitely excluded [103]. Indeed, a cell response to the NP presence was registered after new experiments on pumpkin: a more dense cytoplasm with starch-containing organelle was observed concomitantly with NP aggregates in the cytosol [104].

A further report confirmed significant uptake of NPs (namely, Fe_3O_4 NPs) occurring just in pumpkin, and showed subsequent translocation and accumulation in various tissues, at either the root or leaf level [105]; it is worth mentioning that the plants were grown hydroponically, thus the experimental design was quite different from other works [103, 104]. However, when the trials were replicated with lima bean (*Phaseolus limensis*), any magnetite NP uptake could be registered in any part of the plant [105]; this stresses the need for investigating the largest variety of plant species. Indeed, it was found that ZnO NPs could well enter root cells of ryegrass growing in hydroponic culture, and induced severe growth inhibition [100]. However, little (if any) ZnO NPs were transported from root to shoots, and only a few individual ones could move into the stele and were then transported up. Instead, most of the NPs aggregated on the outer root surface, probably adsorbed to large amount of mucilage secreted by root tip and hairs [100]. By contrast, both NP uptake and translocation were reported to occur in barley seedlings; at least a part of Pd recovered in leaves was due to direct entrance of Pd NPs at the root level and subsequent transport [106, 107]. The same Pd NPs were also able to entry pollen grains [108]. Furthermore, a small-sized species such as *Arabidopsis* was able to take up and translocate NPs within a few days of treatment; the NPs, which were administered simply by watering, followed a pathway from the root cortex to the stele and then to leaves [109]. Research on pumpkin, on the other hand, showed that plants grown in sand and periodical irrigated with the particle suspension exhibited significant decrease in the levels of uptake and accumulation, compared to hydroponic culture system. Evenly, no uptake at all occurred in pumpkin cultivated in soil media, probably due to unavailability of the Fe_3O_4 NPs; presumably, the NPs could have been attached to soil components [105].

It may be surprising that different plant species showed quite different response; it is difficult to find a reason (at the morphological or physiological level) for certain species could take up NPs and some others could not. Indeed, the way the plants were fed with NPs is not well comparable among the reports. Anyway, a number of studies provide convincing evidence that plants, at least some species, are potential pathways of NP transport in the environment. The role of plants in NP bioaccumulation through the food chain should be kept in mind.

Beyond the arena of agronomy and plant pathology, NPs can also be employed for plant transformation purposes. Nearly thirty years ago, foreign DNA was delivered into soybean embryo cells by means of microparticle technique (1 to 5 µm gold particle), obtaining stable transformation [39]. Later on, mesoporous silica NPs with 3-nm pores, capped with functionalized gold NPs, were used to transform protoplasts and plantlets of tobacco, demonstrating that they can serve as an efficient delivery system and make the DNA accessible to the transcription machinery. The mesopores are advantageous in that they can be loaded with a large quantity of chemicals, including those which do not pass membranes or are incompatible with cell culture medium [40]. Carbon nanotubes are able to easily penetrate cell membranes; SWNTs labelled with fluorescein isothiocyanate were internalized by tobacco cells, probably by endocytosis [101]. After 24 h incubation, the cells displayed no apparent damages as for membrane integrity, normal cell morphology and cytoplasm fluidity, while the proliferation rate was comparable to that of untreated cells. Furthermore, also DNA delivery to tobacco cells by means of SWNTS could be performed successfully. It was observed that the two SWNT conjugates, those with fluorescein isothiocyanate or with DNA, had finally different distribution within the cells: the one, just at vacuoles while, the other, in the cytoplasmic strands between the vacuoles. It may be concluded that these results are very promising in the view to effectively deliver any sort of molecules to plant cells, possibly targeting different cell compartments, for various purposes such as cell imaging, cell regulation, or transformation [101].

As for the intimate interaction of carbon nanotubes with cells, SWNTs were identified as a novel class of ion channel blockers [110]; in particular, in Chinese hamster ovary cells, SWNTs block K^+ channel subunits in a dose-dependent manner. Probably SWNTs can achieve this by fitting into the pore and thus hindering ion movement or, alternatively, preventing conformational changes. Indeed, the toxic potential of CNTs is still controversial: both toxic and non-toxic effects are reported, either *in vitro* or *in vivo* [111]. This could depend on the presence, or not, of residual catalyst metals (Co, Ni, Fe, Mo) located inside or outside the CNTs, and of amorphous carbon present in some preparations. ROS production in human lung cells, as elicited by CNTs, was in fact reduced by using CNTs free of metal impurities and residual carbon. Therefore, the importance of experimenting with well-characterized materials has to be underlined [111].

NP PHYTOTOXICITY

Plants are the primary producers, in other words, they are basic components of ecosystems. Both lower and higher plants, native and crop species, have necessarily to do with NPs during any phase of the life cycle, since NPs of various origin (natural; anthropogenic, either unintended formed or engineered) are being continuously released into environments. Detrimental interactions of plants with NPs could not be excluded a priori; indeed, assessment of possible risks is urgent before NPs become more and more ubiquitous in every aspects of life. Presently, understanding phytotoxicity of NPs is still at the beginning [112] compared to larger amount of information on animals.

Lower Plants

Research on NP toxicity to lower plants has focused specifically on algae. Algae are at basis of the food web in all aquatic environments. Furthermore, the ecological role of marine microalgae is also crucial for their photosynthesis evolves about half of the total O_2 production. Green algae are quite heterogeneous from structural point of view. They share the main photosynthetic pigments, that is, chlorophyll *a* and *b*, with higher plants. Unicellular green algae have been considered by a number of studies aimed to identify potential harmful effects of NP pollution to aquatic wildlife environment.

It has to be stated that specific hazards are associated with release of nanoscale products into water systems [113, 114]. Overall, ecotoxicological effects clearly depend on the nature of the particles and on the test organism structure and physiology [115, 116]. Again, it has to be stressed that risk assessment of NPs requires careful, specific approach. For instance, the analysis of the particles prior to and during toxicity assessments has to be emphasized [117, 127]. Surprisingly, it was shown that the efficiency of photosynthesis of the freshwater green alga *Pseudokirchneriella subcapitata*, which is perhaps the most studied microalga within this research area, was not affected by NPs of TiO_2, ZrO_2, Al_2O_3, CeO_2 and polymethylmethacrylate (0.06, 0.41, and 1.08 μm) even at concentrations of 100 mg L^{-1}; this contrasted with other reports. The proposed reason for the unexpected absence of ecotoxic effects, at least according to the authors, was the low bio-active concentration caused by rapid aggregation and/or coagulation of free NPs [118].

The unicellular green alga *Chlamydomonas reinhardtii*, which is easy to be cultured and displays great sensitivity to external stimuli, was used as a model organism to test toxicity of two widely employed NPs, TiO_2 and QDs [53]. QDs affected *Chlamydomonas* more severely compared to TiO_2 NPs. Indeed, cell growth inhibition begun at 1 mg L^{-1} QDs, whereas this same concentration of TiO_2 NPs did not produce effects. Conversely, algal growth was significantly impaired after two days of incubation at 10 mg L^{-1} TiO_2 NPs, and was completely inhibited on day three; it resumed thereafter, and finally (day five) was comparable to controls. At the much higher concentration of 100 mg L^{-1} TiO_2 NPs, a noticeable growth decrease was evident already at day one, and continued for two more days; then, there was a recovery, and the final cell concentration attained 80% of controls. Overall, the growth inhibition was significant, but temporary at any concentration. Cell aggregation was observed at an extent dependent on TiO$_2$ NP concentration; however, also this effect was temporary. At a physiological level, little impact on photosynthesis of to *Chlamydomonas* was observed at 1 to 100 mg L^{-1} TiO$_2$ NPs. NP toxicity to the alga was studied also through physiological/biochemical and molecular approaches [53]. In *Chlamydomonas* cells growing in the presence of TiO$_2$ NPs, levels of malondialdehyde (MDA), which is an indicator of lipid peroxidation, were transiently increased in a dose-dependent manner, with a maximum at 12 h; thereafter, the levels declined. Further insights were given by analysing the transcriptional expression profiling of four stress response genes (*cat*, *gpx*, *sod1*, and *ptox2*) in *Chlamydomonas* cells exposed to NPs [53]. Transcripts of *cat* and *sod1* genes were significantly enhanced at as low as 0.1 mg L^{-1} QDs. As for TiO$_2$ NPs, up-regulation of the antioxidant genes *cat*, *gpx*, and *sod1* occurred only at 1 and 10 mg L^{-1}, and not at 100 mg L^{-1}. The increase in transcript levels of *cat* and *gpx* genes was far more precocious signal of nanotoxicity compared to the alteration of growth kinetics, since it appeared at already 1.5 and 3 h of cell exposure, respectively. However, since environments may offer many sources of oxidative stress, enhanced expression of antioxidant genes could not be assumed as specific biomarker for environmental toxicity of NPs *in vivo*, but it could have only a complementary value [53]. These findings on *Chlamydomonas* are the first evidence of oxidative stress due to NPs occurring in plants; it may be inferred that oxidative damage is a common nanotoxicity response of plant and animal cells. While such a response has been exhaustively described in animals, however, many additional studies are needed to depict a complete scenario for plants.

The microalga *Pseudokirchneriella subcapitata* was also employed in chronic (96 h) toxicity tests with TiO$_2$ NPs of 10 nm in size, resulting more sensitive than the cladocerans *Ceriodaphnia dubia* and *Daphnia magna*, and much more than the freshwater fish, fathead minnow, with IC$_5$ values of 1 and 2 mg L^{-1} [119]. Such an order of sensitivity among these different organisms was consistent with previous observations [120, 121]. Furthermore, a key study finding was that altering test water characteristics can alter TiO$_2$ toxicity; the presence of cladoceran food component, such as an inorganic substrate (kaolinite clay) or, to a greater extent, organic carbon, decreased TiO$_2$ toxicity to *C. dubia*, by decreasing the bioavailability of TiO$_2$ [119].

Primary producers passes biomass to consumers; however, they may also function to promote pollutant transfer through the food web. This was the theme of a study which used *Pseudokirchneriella subcapitata* to describe effects of emerging environmental contaminants such as QDs and understand their fate through trophic levels [121]. Although a lethal endpoint of QDs to *P. subcapitata* was determined (LC$_{50}$: 37.1 ppb at 96 h), the intact nanocrystals in fact allowed the alga to survive at much high cadmium concentrations; in particular, they gave protection up to 75-fold that of the cadmium LC$_{50}$ of 129 µg L^{-1}. QDs entered algal cells, resulting in a change of cell integrity and structure. Moreover, the transfer of QDs from the alga *P. subcapitata* to a primary consumer, *Ceriodaphnia dubia* was also evidenced [121]. The authors hypothesized that potential transfer of toxic core metals to higher trophic levels and - therefore - biomagnification can occur. In addition, if algal uptake results in QD breakdown and core metal release, lethality to primary producers may even interrupt the food chain. By contrast, others failed to observe internalization of SiO$_2$ NPs with 12.5 and 27.0 nm diameter into *P. subcapitata*, notwithstanding the algal growth was inhibited; the particles were seen to be adsorbed to the cell wall, suggesting that toxicity could occur through surface interactions [117].

Besides focusing on inherent toxicity of NPs, it is also important to consider the possible interactions with existing environmental pollutants. This topic was developed by using algae and crustaceans challenged with C$_{60}$-nanoparticles (Buckminster fullerene): these have 60 linked carbon atoms in a cage-like structure similar to a soccer football [115]. It was indicated that fullerenes can cross the cell membrane and localize into cell compartments [122]. The C$_{60}$ were *per se* toxic to the alga *P. subcapitata*, though the uncontrollable NP aggregation caused no reproducible dose-response relationships could be established. Aggregates, in fact, adsorbed on algal cells were seen. Four environmental contaminants were used as model compounds (atrazine, methyl parathion, phenanthrene, and pentachlorophenol, PCP); 85% of phenanthrene sorbed to C$_{60}$-aggregates, whereas only 10% sorption was found for the other pollutants. Consistently, an increase in toxicity was observed in algal growth test, with a decrease of EC_{50} from 720 µg L^{-1} (without C_{60}) to 430 µg L^{-1} phenanthrene due to C_{60} addition [115]. This demonstrates unequivocally that phenanthrene, a known narcotic substance targeting cell membrane, sorbed to C_{60}-aggregates is available for the organisms. As for atrazine, also a tendency for an increase in toxicity, though not significant, was registered. In contrast, the toxicity of PCP to *P. subcapitata* decreased by a factor of 2 in the presence of C_{60} suspensions; toxicity of methyl parathion was unvaried. The other test organism, *Daphnia magna*, was less vulnerable. It produced different results with phenanthrene due to the fact it is provided with a carapace; through moulting, the latter contributed to dispose off the C_{60}-aggregates together with the sorbed phenanthrene. Moreover, both desorption and active excretion from the digestive tract occurred in *Daphnia*. This study is of great importance, in that it demonstrates for the first time the influence of NPs on aquatic toxicity and bioaccumulation of other environmentally relevant contaminants [115]. Again, it has to be stressed that different structure and physiology of the test organisms lead to quite different conclusions.

The important issue of interactions between NPs and other compounds of the aquatic environment was also treated [123]. Test organism was the coastal marine diatom *Thalassiosira weissflogii* which, like as many phytoplankton organisms, produces and secretes polysaccharide-rich exopolymeric (EPs) substances. These are involved in formation of marine gels, marine snow, and biofilms [124] and may provide a large number of strong binding sites for environmental toxicants, for instance trace metals, since they also contain a considerable amount of covalently bound proteins rich in thiol functional groups. Therefore, EPs could have a role in modulating the toxicity of NPs. In particular, 60-70 nm sized-Ag NPs were studied [123]. It was found that toxicity of Ag NPs to the diatom, which was evaluated by means of cell growth, photosynthetic activity, and chlorophyll production was strictly dependent on the extent of silver ion dissolution. The study pointed out that a substantial amount of Ag^+ could be released from Ag NPs even when they were highly aggregated [123]. This is important when referred to the aquatic environment since it could influence the effects of metallic NPs; in particular, Ag^+ is one of the most toxic trace metals known. Indeed, when the free Ag^+ concentration was greatly reduced, no toxicity to *T. weissflogii* was observed. Interestingly, the alga significantly increased the secretion of polysaccharide-rich EPs at increasing the concentration of free Ag^+. This could be interpreted as induction of carbohydrate synthesis under stress conditions, which may help algae build a more suitable, surrounding microenvironment [123]. Carbohydrate extrusion by algae has been previously proposed as detoxification mechanism for Cu^{2+} and Cd^{2+}, based on the fact that the algae which produced the highest EP amounts were those able to tolerate the highest metal concentrations [125].

Toxicity of silver NPs (10 to 200 nm) was also studied in the freshwater alga *Chlamydomonas reinhardtii* [126]. Concern about Ag NPs is due their large use, extended to a huge variety of consumer products (for instance, textiles, paints, food supplements, laundry additives) making it likely they are released to aquatic environment (see section 1). In bacteria, the mechanism of Ag NP toxicity is quite controversial, since it is not clear whether it depends on the NPs *per se* or it is due to Ag^+ ions; for instance, it was established that toxicity to Gram-negative bacteria was related mainly to direct effects of NPs, which were found to be accumulated within the cells and at the cell membrane [127]. Photosynthesis *Chlamydomonas reinhardtii*, was assumed as an endpoint; by referring toxicity data to Ag^+ concentration, it was found that Ag NPs were more toxic than $AgNO_3$, and it could be concluded that Ag NP toxicity to *C. reinhardtii* was mediated by Ag^+ [126]. The study stresses the importance of fully understanding interactions of NPs with living organisms: the alga, in fact, secreted H_2O_2 in the medium, and this resulted in oxidative release of Ag^+ ions from the NPs. In other words, it was the alga itself which, ultimately, modulated Ag NP toxicity, through both assimilation of Ag (which, in turn, enhanced further Ag dissolution) and the metabolic activity which released a very reactive compound, H_2O_2 [126]. The results on *C. reinhardtii*, therefore, evidenced a toxicity mechanism of Ag NPs different from that putatively operating in bacteria [127]. Conversely, it was found that the toxicity of particulate zinc oxide to *Pseudokirchneriella subcapitata* was dependent on the concentration of soluble ZnO [128]. The role of metal dissolution was also evidenced in a study on Cd-Se QD toxicity to liver cells, which was due to the liberation of free Cd^{2+} ions [129].

A relationship between specific NP properties and the diversity of toxic effects was evidenced by studying TiO_2 NPs [116]. Growth inhibition of the green alga *Desmodesmus subspicatus* and immobilization test with *Daphnia magna* were performed in the presence of two different types of TiO_2 NPs (25 and 100 nm), which were illuminated to induce their photocatalytic activity; the latter is induced by UV light, and results in photochemical degradation of substances. This method is used, for instance, for water decontamination or to obtain a self-cleaning (bactericidal) effect on the surface of various materials. The 25 nm-sized particles, but not the larger ones, displayed a concentration-dependent toxicity on algal growth; in contrast, both products were noxious to *Daphnia*, though the dose-effect was less pronounced. Therefore, ecotoxicity strictly depends on the test organisms and their structural organization or physiology [116].

Higher Plants

NPs are being increasingly used in agriculture; the prediction is that nanotechnology will deeply transform both agricultural and food industry. NPs have in fact a great potential to substantially improve agricultural production through several strategies [130]. For instance, nanosensors linked to global positioning system (GPS) could thoroughly monitor in real-time either soil conditions or plant growth in the field; they could therefore identify type and location of problems (heat, drought) allowing farmers to make early and tuned interventions. This is a highly desirable goal in the so-called "precision farming", which is aimed to maximize crop yields while minimising the use of water, fertilizers, and pesticides. Furthermore, nanoscale devices could be used to monitor plant health, providing appropriate, either preventive or curative, remedies [130]. Chemical delivery to crops increasingly relies on nanotechnology; this can be achieved by means of encapsulation of herbicides and pesticides within NPs, from which a controlled release (slow or quick, depending on the necessity) can be obtained. Overall, many developed and developing countries have been significantly funding nanothechnology programmes with a specific focus on agricultural applications [130].

For the mentioned increasing use of NPs in the agricultural sector, and for the fact that plants and plant-derived foods are routinely consumed by humans and many animals, studies on potential toxicity of engineered NPs to various crop species are being more and more stimulated. In the view of introducing NPs into soil, most of the toxicity tests consider seed germination and/or root growth as an endpoint. A few reports testify for positive effects of NPs. Nano-SiO_2 and TiO_2 increased the rate of germination and growth of soybean seedlings and also stimulated nitrate reductase activity [131]. Growth of spinach was enhanced by treating with nano-TiO_2, an effect due to promotion of photosynthetic and nitrogen metabolism [132-134]. A biphasic effect was observed in *Zea mays* plantlets treated with magnetic NPs (8 nm average size) coated with tetramethylammonium hydroxide; low concentrations stimulated plant growth, while high concentrations induced inhibitory or toxic effects; in particular, brown spots on the leaf surface were putatively ascribed to oxidative stress due to iron excess [102]. However, studies evidencing clear negative effects of NPs to plants are being accumulating. A number of crop species (corn, cucumber, soybean, cabbage and carrot) exhibited a significant reduction in root length due to exposure to alumina NPs at 2000 mg L^{-1}. The NPs were either loaded with phenanthrene, or unloaded; in latter case they exhibited higher toxicity [135]. Various types of NPs (MWCNTs, nano-Zn and Al, nano-ZnO and Al_2O_3) were used to study effects on the first life stages of six higher plant species (radish, rape, ryegrass, lettuce, corn and cucumber) [136]. It resulted that seed germination, intended as radicle emergence or cotyledon appearance, was not affected except for corn and ryegrass, which were inhibited by nano-Zn and nano-ZnO, respectively, at 2000 mg L^{-1}. A protective role of seed coats possibly explained the scarce NP effect on germination. In contrast, root elongation tests performed with nano-Zn and nano-ZnO demonstrated dose-dependent inhibition. Fifty percent inhibitory concentrations (IC_{50}) of nano-Zn and nano-ZnO were estimated to be near 50 mg L^{-1} for radish and 20 mg L^{-1} for rape and ryegrass [136]. Therefore, the response largely depended on the plant species. This same conclusion was shared by a pilot study regarding SWCNT toxicity to commonly used crop plants such as cucumber, cabbage, carrot, lettuce, tomato, and onion [137]. The nanotubes, which measured approximately 8 nm in width and from a few hundred nm to a few µm in length, were either functionalized with poly-3-aminobenzenesulfonic acid (fCNTs) or non functionalized (CNTs). The CNTs affected root elongation more than the fCNTs, which produced inhibitory effects only in lettuce, and in one experiment out two. The fact that functionalization resulted in a change in nanotube toxicity agrees with previous findings [135]. In particular, root elongation was enhanced following exposure to CNTs in cucumber (at 24 h) and onion (at 24 and 48 h), whereas root length of lettuce and tomato (at both 24 and 48 h) was negatively affected [137]. Tomato was the most sensitive species. In contrast, cabbage and carrot were unaffected by either form of CNTs [137]. Therefore, it seems to be confirmed that the genetic diversity is a primary factor influencing the plant response. Secondly, the seed size may play a role, rendering a seed more sensitive to NPs: for instance, a large-seeded species such as cucumber would have a lower surface to volume ratio than small-seeded species (*e.g.*, tomato lettuce, onion), for which increased effects should had been expected [137]. Interestingly, the study tried also to assess the possible uptake of CNTs by roots. Cucumber was chosen as a representative plant root: no uptake occurred after 48 h exposure, neither of CNTs nor fCNTs, as shown by SEM observations. According to the authors' conclusion, this does not exclude that using longer exposures, or testing a more sensitive species to CNTs could lead to different results [137]. Nanotubes, either CNTs or fCNTs, were found to be extensively aggregated in sheets on the cucumber root surface, especially adsorbed on secondary roots. The aggregation may have important involvements: it could alter the surface chemistry of the root and, therefore, plant-microbe interactions in the rhizosphere, or essential biochemical processes for mineral nutrition and therefore, ultimately, for optimal plant growth [137].

Overall, a tentative conclusion was that the use of CNTs in agriculture seems not to be adverse [137]. However, the authors themselves admit that further studies are needed to better support what their preliminary results suggested. They simply tested effects on seed germination under aqueous conditions; instead, more realistic exposures (such as soil systems), and a larger variety of species with prolonged exposure periods until more advanced developmental stages have still to be investigated. For instance, several adverse effects produced by ZnO NPs on *Lolium perenne* (ryegrass) were shown [138]: in the presence of the NPs, the seedling biomass significantly reduced, root tips shrank, tegument and cortical cells highly vacuolated or collapsed, and even endodermal and vascular cells impaired. This implies that NPs were able to pass from one cell to another either *via* apoplast (that is, through cell walls) or *via* symplast, namely cell-to-cell by means of plasmodesmata, which are cylindrical cytoplasm channels with 40-60 nm in diameter; thus, large enough to allow the NP passage. Overall, root growth of ryegrass was severely inhibited by ZnO NPs [138].

To overcome the problem of water insolubility of most NPs, an appropriate experimental protocol which avoided NP precipitation and ensured a homogeneous distribution within the culture medium, that is, plant agar media was adopted [139]. Two important crop plants, mung bean and wheat, were studied for their growth response to Cu NPs. After an exposure period of 48 h to concentrations from 200 to 1000 mg L^{-1}, adverse effects on both species were observed. In particular, mung bean was more sensitive than wheat, with median effective concentrations (EC_{50}) for seedling growth of 335 and 570 mg L^{-1}, respectively. Root growth was a more sensitive endpoint than shoot growth, suggesting that only a small portion of NPs could be transported within the plant and that the site of toxic action might be the root. Although cupric ion was solubilized from the Cu NPs, its concentration was not high enough to be toxic; indeed, the toxic effect was caused by the Cu NPs directly. Ultrastructural analysis revealed that the NPs were clearly present in the cytoplasm and cell wall of root tissues, with larger agglomerates at increasing NP concentrations. Furthermore, it was evaluated that bioaccumulation of NPs increased with increasing NP concentrations in growth media. Due to different root apparatus morphology, provided with numerous thin roots, greater accumulation of Cu NPs was found in wheat compared to mung bean. It was concluded that concerns about the toxicity of NPs need to be addressed, and special care should be taken in the application of NPs to various products: soil ecotoxicity and bioavailability to plants have been clearly demonstrated by their research [139].

Phytotoxicity of four rare earth oxide NPs (nano-CeO_2, nano-La_2O_3, nano-Gd_2O_3 and nano-Yb_2O_3) was investigated on several crop plant species (radish, rape, tomato, lettuce, wheat, cabbage, and cucumber), representing a broad range of genetic diversity, different seed size and type of seed coats. NPs may or may not affect the germination, depending on pores at the surface of teguments which allow the NPs pass trough [140]. Consistent to previous observations, the effects on root growth varied greatly between different NPs and plant species, and different IC_{50} were produced. A suspension of 2000 mg L^{-1} nano-CeO_2 (the only tetravalent rare earth oxide tested) had no effect on the six species, except lettuce, whereas the same concentration of other NPs severely inhibited root elongation in all cases. Furthermore, a dose-response phenomenon called 'hormesis effect' could be observed with a low-dose stimulation and a high-dose inhibition. Interestingly, the inhibition concerned different developmental stages in different species; for instance, wheat was affected during seed incubation, while lettuce and rape during both seed soaking and incubation. The authors claim the toxic effects were at only negligible extent due to ions released from the NPs. However, they did not verify whether the NPs were present within the plant tissues. Indeed, the NPs adsorbed on the root surface, due to large amounts of mucilage, might have been dissolved and transported into the root [140].

Near to the numerous morphological evidences, a very few reports provide a deeper inside view of NP phytotoxicity at the cytological or physiological level. Magnetic NPs were able to induce cytogenetical changes such as chromosomal aberrations and perturbation of cell division in 2-3 day-old plants of maize [141]. Genotoxic impact of silver NPs in plant cells was described in detail by using a validated bioassay, that is, the *Allium cepa* root tip cells; it was shown that the NPs decreased the mitotic index in a dose-dependent manner [142]. The study also identified different kinds of chromosomal aberrations were observed with different NP concentrations: for instance, chromosomal stickiness, which is a common sign of toxic influence on chromosomes. Disturbed metaphase, a symptom of disturbance in the spindle apparatus, and cell wall disintegration were also registered in *A. cepa* cells. A further study on magnetic NPs focused on effects to photosynthetic pigments and nucleic acids of *Zea mays* plantlets [102]. The chlorophyll *a* level was increased in the presence of small ferrofluid concentrations, while was lowered (about 35% decrease) at enhanced concentrations. The same was observed for chlorophyll *b* and total carotenoid content. Overall, the efficiency of photosynthesis as indicated by chl. *a*/chl. *b* ratio was reduced for all the concentrations. The average nucleic acid level was slightly enhanced compared to controls. In addition, the authors hypothesized that the NPs they used could influenced locally the transmembrane flows through ion channels due to their magnetic properties [102].

By supplying a colloidal suspension of inorganic nanoparticulate material of natural (bentonite clay) or industrial (TiO_2) origin, interference with water transport in *Zea mays* was documented [143]. Indeed, NP suspensions had flow inhibitory effects on excised roots, which appeared to be concentration dependent and progressive. A presumably physical interaction between colloidal particles and root cell walls appeared to be involved in inhibiting root hydraulic conductivity, in particular, reducing cell wall pore diameter. When NPs were given to intact seedlings by addition to hydroponic solution, rapid inhibition of leaf growth and significant reduction of transpiration was registered: the leaf response and the reduced root water capacity due to NPs are evidently related. By contrast, irrigation of soil-grown plants with the same NP suspension had mostly insignificant inhibitory effects on long-term shoot production. This is the first report to establish that colloidal suspensions of nano-particulate materials in the root media can reduce the hydraulic conductivity of maize primary roots and induce symptoms of water stress in the seedling shoots. These symptomatic responses were rapidly initiated and appeared to involve physical rather than toxic interactions between nanoparticles and roots. The study points out that the physical effects of NPs should also be considered when evaluating their potential environmental impacts on plants. For example, crop irrigation in drought-prone regions of the world will increasingly employ waste recycled waters containing high concentrations of suspended colloids and dissolved biopolymers, or waters contaminated with industrial NP wastes. It is conceivable that this will cause significant root clogging and thereby limit root water uptake under field conditions [143].

TOXICITY OF PD NPS TO THE DIPLOID AND HAPLOID PLANT GENERATION

The new automotive exhaust catalysts involve noble metal emissions into the environment. Among the noble metals, Pd has the greatest mobility. Pd was in fact recovered even in a depth of 12-16 cm in the ground near a German highway [19]. Moreover, Pd uptake by plants is also greater than that of Pt and Rh; its effects to cells are also more dangerous [144]. Pd is one of the main catalytic elements and about 90% of the Pd emissions are particulate matter, prompted to investigate biological effects of such a material that is increasingly accumulating in the environment. To this aim, well characterized Pd particulate material, uncontaminated with other metals or various Pd transformation products, was expressly prepared (Fig. **3**); on a best possible level, it could resemble that emitted from automobile catalysts into the environment [145]. Though exposure experiments with original road dust might appear more environmentally relevant, however, they suffer the problem that a lot of other traffic-emitted metals such as Pb, Zn, Cd, Sb, and Cu will surely overlap with Pd effects, therefore making it no possible to understand clear biological responses to Pd.

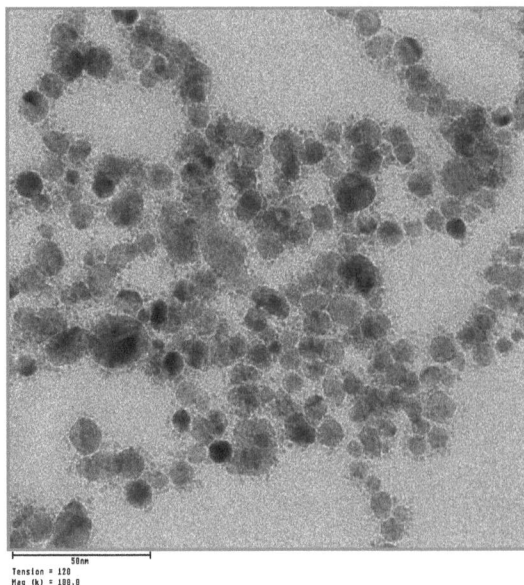

Figure 3: TEM image of expressly synthesized Pd NPs, which resemble those emitted from new automotive exhaust catalysts [145].

The expressly synthesized Pd NPs were tested on both vegetative and reproductive plant systems: in particular, barley seedlings [106, 107] as a representative of diploid, sporophytic apparatus, and kiwifruit pollen [108] as a representative of haploid, gametophytic portions, respectively. Differently from animals, in fact, two distinct individuals or generations alternate each other in the higher plant life cycle; one, which is responsible for all vegetative functions and the other only devoted to sexual reproduction.

Barley seedlings were grown on nutrient solutions with either smaller (1-12 nm) or larger (~ 1 μm) Pd NPs; the small series resembled the particle emitted from exhaust gas catalytic converters. After a week, significant reduction in leaf length was observed in plants exposed to the small particles; furthermore, stress symptoms appeared at the leaf level with increasing Pd concentrations [106]. The alterations consisted in leaf rigidity and convolution. By contrast, neither growth inhibitory effects nor leaf stress symptoms were registered in seedlings treated with the larger particles; leaf length was somewhat reduced only at Pd concentrations ≥ 40 μmol·L^{-1}.

Moreover, the extent of Pd recovered into exposed plant leaves largely depended on the particle size: it was 5 to 15-fold higher in plants exposed to the smaller particles, compared to plants exposed to the greater ones. In particular, the Pd uptake from the small particles increased disproportionately to increasing Pd in the culture medium, probably indicating that the plant defence system was largely dismantled. Furthermore, the lower Pd uptake from the larger particles correlated well with absence of stress symptoms. These findings provide clear evidence that barley roots are able to take up Pd, and that translocation to leaves can then occur *via* xylem vessels. As for the form of Pd taken up, the authors claim that uptake of either the Pd NPs directly or dissolved Pd(II) could have taken place. Root exudates might also have influenced the Pd dissolution. It is great worth of this study the demonstration that even low levels of particulate Pd, which resembled that due to traffic related emissions, can cause detrimental effects on vegetative plant development [106].

Pd NPs entrapped in an aluminium hydroxide matrix, and other metal NPs, all from commercial source, were tested on lettuce seed germination [146]. No statistically significant influence on plant growth resulted when the seeds were planted immediately after adding the NPs to soil. In contrast, Pd and Au NPs at low concentrations, Si and Cu NPs at higher concentration, and a combination of Au and Cu NPs significantly increased the shoot/root ratio compared to controls when seeds were planted after 15 days of incubation of the NPs within the soil. Thus, the NPs may not have exerted a direct, but indirect, influence on plant growth. Worth of this study is the attention it brought on various problems relevant to ecotoxicity of NPs, such as the influence on the soil microbial community. Even though no significant effects on soil diversity over a short term could be noted, indirect or longer term effects could be not excluded [146].

Further work on barley as affected by Pd considered three different types of Pd particles: Pd NPs dispersed on micrometer silica support particles (Pd/SiO$_2$), Pd-only NPs (Pd NPs), and Pd micrometer particles (Pd MPs) [107]. Barley seed were exposed to nutrient liquid medium containing different amounts of the three particles, and the resulting plant development assessed after 2 weeks. Leaf length was linearly decreased after Pd NP and Pd MP exposure, while no significant correlation was observed in the Pd/SiO$_2$ treatment (Fig. **4**). Pd uptake and compartmentalization into different plant segments were evaluated. Linear correlations between Pd exposure concentration and Pd content in roots and seeds were observed, whereas Pd uptake followed polynomial regression in leaves. In any plant portion, the highest uptake was registered in the Pd NP treatment. Importantly, the Pd NPs were recovered in the sap of barley seedling by TEM analysis (Fig. **5**), confirming the previous hypothesis of direct NP uptake *via* roots.

Figure 4: Barley seeds and plantlets exposed to different types of Pd particles. (Images kindly provided by Dr. F. Battke, Institute of Biochemical Plant Pathology, Research Center for Health and Environment, Helmholtz Center Munich, Germany.)

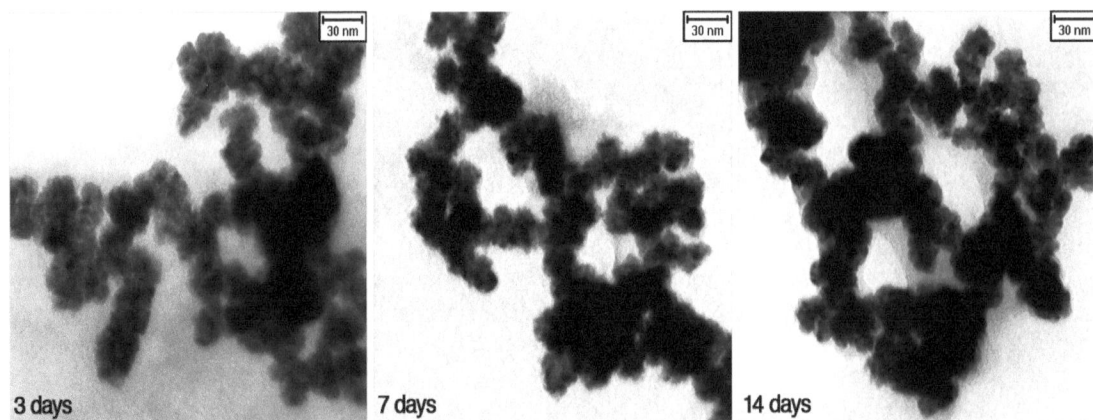

Figure 5: TEM image of Pd NPs recovered in barley seedlings sap at various days of treatment.

Moreover, levels of potassium and calcium, which are representatives of important macroelements, followed a significant, logarithmic decrease with increasing Pd content in the medium, in all the three NP treatments. However, strongest affects on plant growth and macronutrient uptake were caused by Pd uptake from Pd MPs and Pd/SiO$_2$. Obviously, Pd NPs can enter the plant leaves to a high amount, but do not affect leaf growth or nutrient uptake as strong as soluble Pd species. However, in this study plants were grown for only two weeks and dissolution of Pd NPs in the plants seems likely, *i.e.,* the high amounts of Pd taken up by the plants will undoubtedly show delayed effects at more advanced developmental stages [107]. The study further stresses the suitability of the particulate Pd matter for exposure studies with plants, showing reasonable and reproducible results.

In the higher plant life cycle, pollen represents the male gametophyte, *i.e.,* the haploid organism that produces the male gametes. Angiosperm pollen is highly reduced, bi-or three-celled (Fig. **6a-c**). The unique pollen task in its short life is to germinate, that is, to produce a fast growing, cylindrical cell prolongation, the pollen tube (Fig. **6d, e, h**), which delivers the male gametes to the female partner for fertilization to occur. A major proportion of crop plant yield consists in seeds and fruits: pollen function is the foundation of the success of sexual plant reproduction. Furthermore, in natural ecosystems as well, pollen function is strictly needed to ensure conservation of genetic variability of native plant species, and thus their adaptive chances. A very active machinery of cell metabolism and a great variety of cell structures are at the base of both the germination and tube elongation processes; therefore, these are extremely sensitive to external and internal factors, and can be affected in the presence of a wide range of chemicals (Fig. **6f, g, i**).

Dose-response curves can be generated; quantification of pollen tube emergence and growth allows the inhibitory effect to be expressed by EC$_{50}$, *i.e.,* the concentration of a test compound that reduces growth to 50% of controls. Indeed, *in vitro* culture of pollen grains provided a highly sensitive indication of potential environmental and pharmaceutical risk [45, 147]. Evenly, pollen tube growth-based tests were proven to be at least as sensitive and reliable an indicator for detecting basal toxicity as mammalian cell cultures [148].

In vitro performance of kiwifruit pollen was deeply influenced by 5-10 nm sized Pd NPs [108]. A number of biological effects, including morphological alteration of the grains, damage to cell membranes, and depletion of endogenous calcium were observed. Finally, this resulted in dramatic inhibition of pollen tube emergence and growth, with EC$_{50}$ of 0.23 and 0.13 mg L^{-1} at 2 h, respectively. Interestingly, Pd NP concentrations which produced significant inhibition on pollen performance were comparable to those found in dusts along high traffic roads [18]; indeed, the concentrations effective on pollen function were markedly lower than those which generated stress effects and reduced leaf length in barley [106]. This seems to further confirm previous observations referring to other chemical stresses; that is, pollen itself, its developmental processes, and the complex morphogenetic event of germination share the peculiar feature of greater sensitivity compared to vegetative plant portions [149-151].

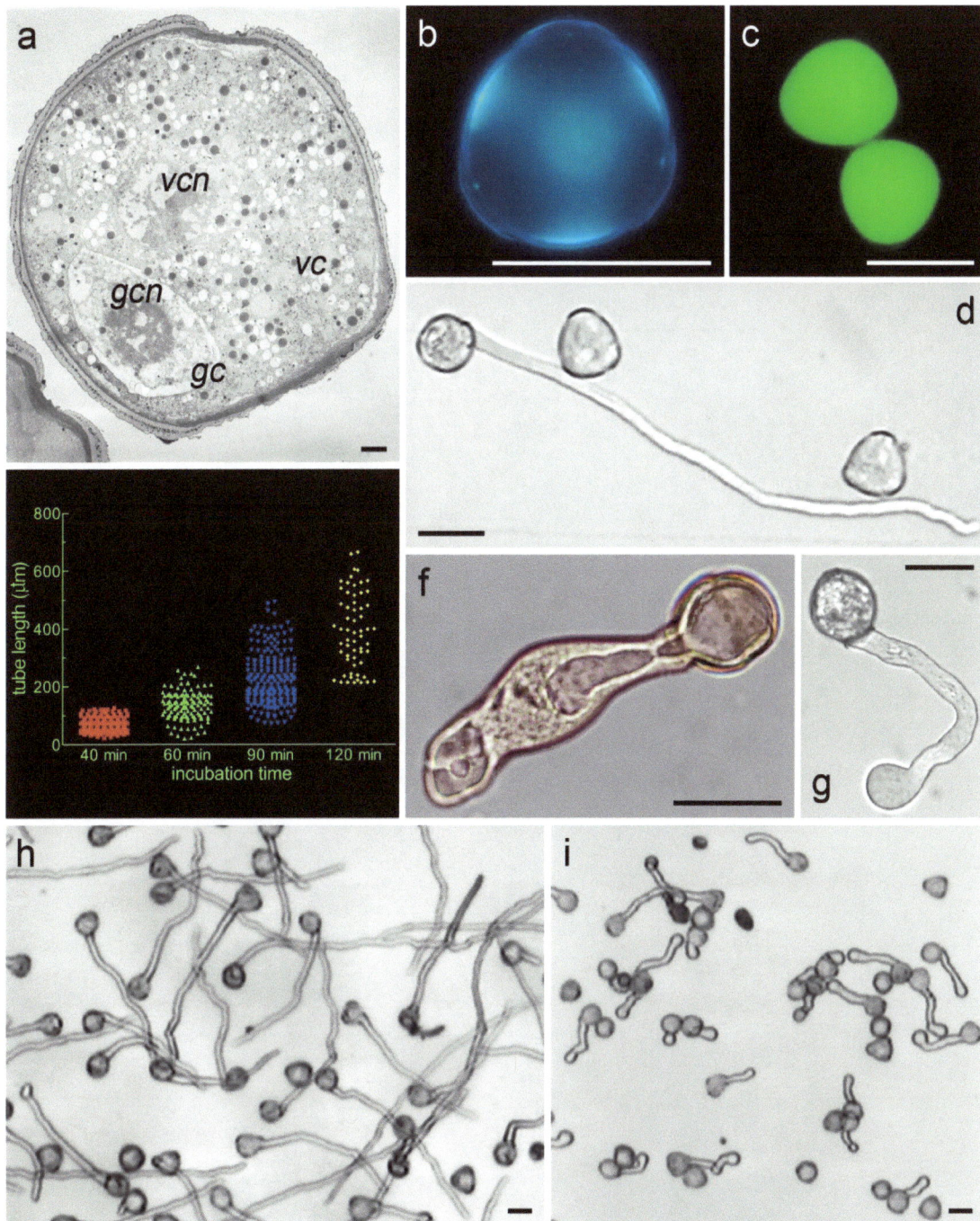

Figure 6: Kiwifruit pollen grains and tubes. **a**: TEM image of ungerminated pollen grain. Kiwifruit pollen is bicellular: the larger vegetative cell (*vc*) contains the generative cell (*gc*), which will produce the male gametes after germination. Vegetative cell nucleus, *vcn*. Generative cell nucleus, *gcn*. **b, c**: fluorescence images of pollen grains treated with decolourised aniline blue to stain callose, and with fluorescein diacetate to test viability, respectively. **d**: scattering of overtime elongation of pollen tubes growing in basal medium. Germination occurs in a scalar manner: individual pollen grains of the population has different vigory, that is, they exhibit different rate of tube emergence and growth **e, h**: light microscopy images of pollen tubes after 2 h incubation in basal medium. **f, g, i**: morphological alteration to pollen tubes and evident inhibition on pollen tube emergence and elongation is induced by a huge variety of chemicals: for instance, metabolic inhibitors such as β-D-glucosyl Yariv reagent (f), which binds and precipitates arabinogalactan proteins, and heavy metals such as chromium (g, i). Bars represent 2 μm (a); 30 μm (b, c, f-i). (a: original image kindly provided by A.R. Taddei; b-i: original microphotograps and graph (e) by A. Speranza).

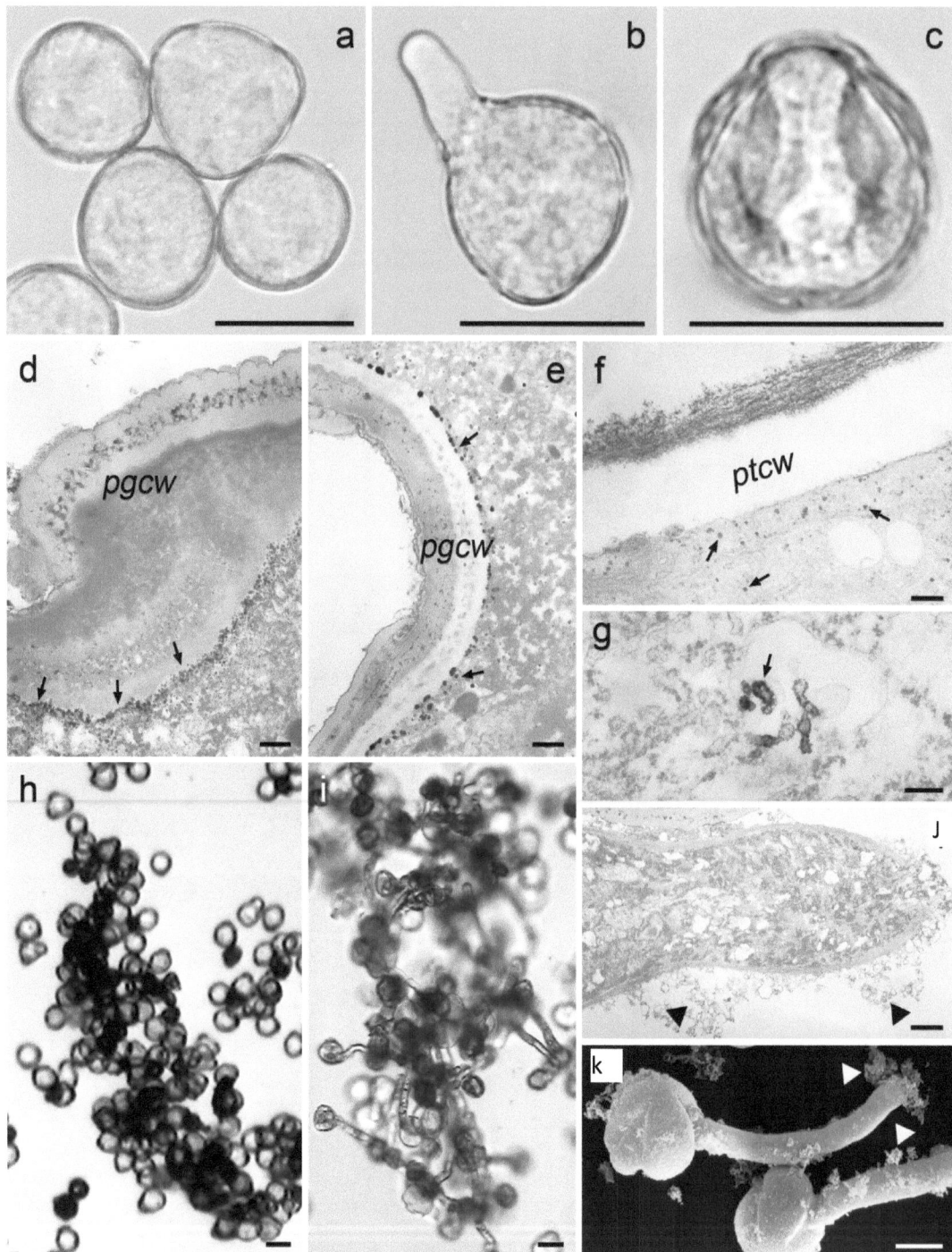

Figure 7: Kiwifruit pollen grains and tubes as spiked with Pd NPs. **a, b**: control pollen grains, ungerminated and germinating, respectively. **c**: shrunk morphology of pollen grain treated with Pd NPs. **d, e**: Pd NPs were recovered as soon as after 15 min exposure, namely at the plasmalemma region in the correspondence of germination apertures (arrows). Later on, they were internalized (**g**). They were found also in tube cytoplasm (**f**). **h, i**: light microscopy observation of grain and tube aggregation as induced by Pd NPs. **l, m**: TEM and SEM image, respectively, of pollen tubes exposed to Pd NPs during elongation phase; leaching of cell material is appearing (arrowheads). *ga*: germination aperture; *pgcw*: pollen grain cell wall; *ptcw*: pollen tube cell wall. Bars represent 30 μm (a-c, h, i); 500 nm (d, e); 200 nm (f, g); 1 μm (l); 10 μm (m). (a-c, h, i: original microphotograps by A. Speranza; **d-g, j, k**: original images kindly provided by A.R. Taddei).

Furthermore, similarly to barley [106, 107], Pd toxicity to kiwifruit pollen was in relation to speciation and the particulate form of the metal: Pd^0 from the NPs was far more dangerous than soluble Pd^{II}, as for *in vitro* performance and lethality as an endpoint, or the extent of Pd uptake and calcium loss. It is worth noticing that the Pd NPs directly and rapidly entered the grains, just as they were [108]. Among any other plant cells, pollen cell wall exhibits the greatest complexity, both structurally and chemically; nevertheless, the NPs could find their way at the level of germination apertures, which interrupt the thick and resistant outer wall (exine). At the beginning, the NPs were found to have accumulated beneath the plasmalemma just in the correspondence of germination apertures; after that, they were recovered also deeper in the cytoplasm of either grains or tubes (Fig. **7d-g**). Under impact with Pd NPs, increasing percentage of kiwifruit pollen grains assumed a shrunk shape (Fig. **7a-c**) in a dose-dependent manner. This did not occur with $PdCl_2$ [108]. Moreover, cell aggregation was observed in kiwifruit pollen after treatment with Pd NPs either at the beginning or during tube growth (Fig. **7h, i**). Aggregation was also described in cultures of *Chlamydomonas reinhardtii* and bacteria exposed to NPs [53, 152]. Aggregation of cells might be related to ability of NPs to form protein-based aggregates [153]. At least in the case of pollen, it is well known that proteins are actively released since the beginning of germination as a 'physiological' pollen feature; furthermore, in the case of kiwifruit pollen, the membrane rupture due to NPs could have induced substantial leaching from cells (Fig. **7j, k**), and this could ultimately result in the formation of a 'glue' linking together pollen grains and tubes [108].

Plants are organisms fixed to their substrate; therefore, differently from animals, they cannot escape pollution and may well experiment cumulative exposure to traffic related emissions, especially when living near to highways and roads with heavy traffic [154]. Therefore, studies with plants give biological responses highly suitable to better elucidate processes and effects of Pd NP ecotoxicity.

CONCLUDING REMARKS

Studying potential NP toxicity to plants include to consider the route of uptake, intracellular accumulation, cellular sites and modes of action, damage to cell structures, and kinetics of toxification. Only a few of these issues have been exhaustively investigated yet in plants, contrasting with the larger knowledge which has been accumulating on animals and humans. On the other hand, NP technology offers a large array of applications also in plants, for either plant biology research or management purposes; in ecosystems, native plant species are inevitably subjected to NP pollution. It is not a paradox that, by improving the information on plants, also the knowledge on risk assessments for human health will be increased as well. This is due to strict dependence on plants of all living beings.

ACKNOWLEDGEMENTS

The authors are grateful to Dr. Marianne Hanzlik (Institute of Electron Microscopy, Department of Chemistry, Technical University Munich, Germany) for TEM images of Pd NPs. Furthermore, Dr. Anna Rita Taddei (Centro Interdipartimentale Microscopia Elettronica, Università della Tuscia, Viterbo, Italy) is gratefully acknowledged for TEM and SEM images of kiwifruit pollen.

REFERENCES

[1]   Ball P. Natural strategies for the molecular engineer. Nanotechnology 2002; 13: 15-28.

[2]   Roco MC. Broader societal issue of nanotechnology. J Nanop Res 2003; 5: 181-189.

[3]   Kaur S Nieuwenhuijsen MJ, Colvile RN. Fine particulate matter and carbon monoxide exposure concentrations in urban street transport microenvironments. Atmos Environ 2007; 41: 4781-4810.

[4]   Pöschl U. Atmospheric Aerosols: Composition, Transformation, Climate and Health Effects. Angew Chem Int Ed 2005; 44: 7520-7540.

[5]   Nowack B, Bucheli TD. Occurrence, behaviour and effects of nanoparticles in the environment. Environ Pollut 2007; 150: 5-22.

[6]   Dahl A, Gharibi A, Swietlicki E, *et al.* Traffic-generated emissions of ultrafine particles from pavement-tire interface. Atmos Environ 2006; 40: 1314-1323.

[7]   Kaegi R, Ulrich A, Sinnet B, *et al.* Synthetic TiO$_2$ nanoparticle emission from exterior facades into the aquatic environment. Environ Pollut 2008; 156: 233-239.

[8]   Roubicek V, Raclavska H, Juchelkova D, Filip P. Wear and environmental aspects of composite materials for automotive braking industry. Wear 2008; 265: 167-175.

[9]   Artelt S, Kock H, Koenig HP, Levsen K, Rosner G. Engine dynamometer experiments: platinum emissions from differently aged three-way catalytic converters. Atmos Environ 1999; 33: 3559-3567.

[10]   Abthoff J, Zahn W, Loose G, Hirschmann A. Serial use of palladium for three-way-catalysts with high performance. Motortech Z 1994; 55: 292.

[11]   Herz KR Shinouskis EJ. Application of High-Resolution Analytical Electron Microscopy to the Analysis of Automotive Catalysts. Ind Eng Chem Prod Rev Dev 1985; 24: 6-10.

[12]   Inacker O, Malessa R. Experimentalstudie zum Austrag von Platin aus Automobilabgaskatalysatoren. In: Forschungsberichte - Bundesministerium für Forschung und Technologie (Germany). Edelmetall-Emissionen, Final Report. Berlin: BMBF 1996; pp. 48-53.

[13]   Gomez B, Palacios MA, Gomez M, *et al.* Levels and risk assessment for humans and ecosystems of platinum-group elements in the airborne particles and road dust of some European cities. Sci Tot Environ 2002; 299: 1-19.

[14]   Kanitsar K, Koellensperger G, Hann S, Limbeck A, Puxbaum H, Stingeder G. Determination of Pt, Pd and Rh by inductively coupled plasma sector field mass spectrometry (ICP-SFMS) in size-classified urban aerosol samples. J Anal At Spectrom 2003, 18: 239-246.

[15]   Ravindra K, Bencs L, Van Grieken R. Platinum group elements in the environment and their health risk. Sci Tot Environ 2004; 318: 1-43.

[16]   Ek KH, Morrison GM, Rauch S. Environmental routes for platinum group elements to biological materials - A review. Sci Tot Environ 2004: 334-335: 21-38.

[17]   Leopold K, Maier M, Weber S, Schuster M. Long-term study of palladium in road tunnel dust and sewage sludge ash. Environ Pollut 2008; 156: 341-347.

[18]   Zereini F, Wiseman C, Püttmann W. Changes in palladium, platinum and rhodium concentrations and their spatial distribution in soils along a major highway in Germany from 1994 to 2004. Environ Sci Technol 2007; 41: 451-456.

[19]   Harrison P, (*ed.*) Emerging challenges: Nanotechnology and the environment. In: United Nations Environment Programme (Kenya). GEO Year Book 2007: An overview of our changing environment. Nairobi: United Nations Environment Programme 2007, pp. 61-68.

[20]   Roco MC. Environmentally responsible development of nanotechnology. Environ Sci Technol 2005; 39: 106A-112A.

[21]   A database of nanotechnology in commercial products; The Project on Emerging Nanotechnologies, Woodrow Wilson Center. Available from: http://www.nanotechproject.org/inventories/consumer/ [Cited: 11th Feb 2010].

[22]   Maynard AD. Nanotechnology: A Research Strategy for Addressing Risk. Washington: Woodrow Wilson International Center for Scholars 2006.

[23]   Richardson SD. Water Analysis: Emerging Contaminants and Current Issues. Anal Chem 2009; 81: 4645-4677.

[24]   Zhang WX. Nanoscale iron particles for environmental remediation: An overview. J. Nanopart Res 2003; 5: 323-332.

[25]   Rittner MN. Market analysis of nanostructured materials. Am Ceram Soc Bull 2002; 81: 33-36.

[26]   Aitken RJ, Chaudhry MQ, Boxall ABA, Hull M. Manufacture and use of nanomaterials: current status in the UK and global trends. Occup Med 2006; 56: 300-306.

[27]   Aldana J, Wang YA, Peng X. Photochemical instability of CdSeb nanocrystals coated by hydrophobic thiols. J Am Chem Soc 2001; 123: 8844-8850.

[28]   Benn TM, Westerhoff P. Nanoparticle silver released into water from commercially available sock fabrics. Environ Sci Technol 2008; 42: 4133-4139.

[29]   Neal AL. What can be inferred from bacterium-nanoparticle interactions about the potential consequences of environmental exposure to nanoparticles? Ecotoxicology 2008; 17: 362-371.

[30]   Blaser SA, Scheringer M, MacLeod M, Hungerbuehler K. Estimation of cumulative aquatic exposure and risk due to silver: contribution of nano-functionalized plastics and textiles. Sci Tot Environ 2008; 390: 396-409.

[31]   Oberdörster G, Oberdörster E, Oberdörster J. Nanotoxicology: an emerging discipline evolving from studies of ultrafine particles. Environ Health Perspect 2005; 113: 823-839.

[32]   Alvarez PJJ, Colvin V, Lead J, Stone V. Research priorities to advance eco- responsible nanotechnology. *ACSNANO* 2009; 3: 1616-1619.

[33]   Environmental Protection Agency (US). Nanotechnology White Paper; Report EPA 100/B-07/001. Washington: The EPA 2007.

[34]   Commission of the European Communities (EU). Communication from the commission to the Council, the European Parliament and the economic and social committee: Action Plan "Nanosciences and nanotechnologies: An action plan for Europe 2005-2009". Brussels: The EU commission 2005, No. 243.

[35]   Powers KW, Brown SC, Krishna VB, Wasdo SC, Moudgil BM, Roberts SM. Research Strategies for Safety Evaluation of Nanomaterials. Part VI. Characterization of nanoscale particles for toxicological evaluation. Toxicol Sci 2006; 90: 296-303.

[36]   Christian P, Von der Kammer F, Baalousha M, Hofmann T. Nanoparticles: structure, properties, preparation and behaviour in environmental media. Ecotoxicology 2008; 17: 326-343.

[37]   Ju-Nam Y, Lead JR. Manufactured nanoparticles: An overview of their chemistry, interactions and potential environmental implications. Sci Tot Environ 2008; 400: 396-414.

[38]   Baumann T Fruhstorfer P, Klein T, Niessner R. Colloid and heavy metal transport at landfill sites in direct contact with groundwater. Water Res 2006; 40: 2776-2786.

[39]   Christou P, McCabe DE, Swain WF. Stable transformation of soybean callus by DNA-coated gold particles. Plant Physiol 1988; 87: 671-674.

[40]   Torney F, Trewyn BG, Lin VS-Y, Wang K. Mesoporous silica nanoparticles deliver DNA and chemicals into plants. Nature Nanotechnol 2007; 2: 295-300.

[41]   Yoo JS. Selective gas-phase oxidation at oxide nanoparticles on microporous materials. Catal Today 1998; 41: 409-432.

[42]   Sau TK, Pal A, Pal T. Size regime dependent catalysis by gold nanoparticles for the reduction of eosin. J Phys Chem B 2001; 105: 9266-9272.

[43]   Trindad, T, O'Brien P, Pickett N. Nanocrystalline semiconductors: synthesis, properties, and perspectives. Chem Mater 2001; 13: 3843-3858.

[44]   Nel A, Xia T, Mädler L, Li N. Toxic potential of materials at the nanolevel. Science 2006; 311: 622-627.

[45]   Kristen U. Main features of basal toxicity: sites of toxic actions and interactions in the pollen tube cell. ATLA-Altern Lab Anim 1996; 24: 429-434.

[46]   Xu Z, Liu XW, Ma YS, Gao HW. Interaction of nano-TiO_2 with lysozyme: insights into the enzyme toxicity of nanosized particles. Environ Sci Pollut Res 2010; 17: 798-806.

[47]   Jones CF, Grainger DW. *In vitro* assessments of nanomaterial toxicity. Adv Drug Deliv Rev 2009; 61: 438-456.

[48]   Long TC, Tajuba J, Sama P, *et al.* Nanosize titanium dioxide stimulates reactive oxygen species in brain microglia and damages neurons *in vitro.* Environ Health Perspect 2007; 115: 1631-1637.

[49]   Oberdöster G. Manufactured nanomaterials (fullerenes, C_{60}) induce oxidative stress in the brain of juvenile large mouth bass. Environ Health Perspect 2004; 112: 1058-1062.

[50]   Prahalad AK, Soukup JM, Inmon, J, *et al.* Ambient airparticles: effects on cellular oxidant radical generation in relation to particulate elemental chemistry. Toxicol Appl Pharmacol 1999; 158: 81-91.

[51]   Veranth JM, Reilly CA, Veranth MM, *et al.* Inflammatory cytokines and cell death in BEAS-2B lung cells treated with soil dust, lipopolysaccharide, and surface-modified particles. Toxicol Sci 2004; 82: 88-96.

[52]   Nel A, Xia T, Mädler L, Lin N. Toxic potential of materials at the nanolevel. Science 2006; 311: 622-627.

[53]   Wang J, Zhang X, Chen Y, Sommerfeld M, Hu Q. Toxicity assessment of manufactured nanomaterials using the unicellular green alga *Chlamydomonas reinhardtii.* Chemosphere 2008; 73: 1121-1128.

[54]   Møller P, Jacobsen NR, Folkmann JK, *et al.* Role of oxidative damage in toxicity of particulates. Free Radical Res 2010; 44: 1-46.

[55]   Adams LK, Lyon D, Alvarez PJJ. Comparative eco-toxicity of nanoscale TiO_2, SiO_2, and ZnO in water suspensions. Water Res 2006; 40: 3527-3532.

[56]   Dellinger B, Pryor WA, Cueto R, *et al.* Role of free radicals in the toxicity of airborne fine particulate matter. Chem Res Toxicol 2001; 14: 1371-1377.

[57]   de Kok TM, Hogervorst JG, Kleinjans JC, Briede JJ. Radicals in the church. Eur Respir J 2004; 24: 1069-1070.

[58]   Shi T, Duffin R, Borm PJ, Li H, Weishaupt C, Schins RP. Hydroxyl-radical-dependent DNA damage by ambient particulate matter from contrasting sampling locations. Environ Res 2006; 101: 18-24.

[59]   Folkmann, JK, Risom L, Jacobsen NR, Wallin H, Loft S, Møller P. Oxidatively damaged DNA in rats exposed by oral gavage to C_{60} fullerenes and single-walled carbon nanotubes. Environ Health Perspect 2009; 117: 703-708.

[60]   Markovich Z, Todorovic-Markovich B, Kleut D, *et al.* The mechanism of cell-damaging reactive oxygen generation by colloidal fullerene. Biomaterials 2007; 28: 5437-5448.

[61]   Fenoglio I, Tomatis M, Lison D, *et al.* Reactivity of carbon nanotubes: free radical generation or scavenging activity? Free Radical Bio Med 2006; 40: 1227-1233.

[62] Baulig A, Garlatti M, Bonvallot V, *et al.* A. Involvement of reactive oxygen species in the metabolic pathways triggered by diesel exhaust particles in human airway epithelial cells. Am J Physiol Lung Cell Mol Physiol 2003; 285: L671-L679.

[63] Foucaud L, Wilson MR, Brown DM, Stone V. Measurement of reactive species production by nanoparticles prepared in biologically relevant media. Toxicol Lett 2007; 174: 1-9.

[64] Helfenstein M, Miragoli M, Rohr S, *et al.* Effects of combustion derived ultrafine particles and manufactured nanoparticles on heart cells *in vitro.* Toxicology 2008; 253: 70-78.

[65] L'Azou B, Jorly J, On D, *et al. In vitro* effects of nanoparticles on renal cells. Part Fibre Toxicol 2008; 5: 22.

[66] Li Z, Hyseni X, Carter JD, Soukup JM, Dailey LA, Huang YC. Pollutant particles enhanced H_2O_2 production from NAD(P)H oxidase and mitochondria in human pulmonary artery endothelial cells. Am J Physiol Cell Physiol 2006; 291: C357-C365.

[67] Pacurari M, Yin XJ, Zhao J, *et al.* Raw single-wall carbon nanotubes induce oxidative stress and activate MAPKs, AP-1, NF-kappaB, and Akt in normal and malignant human mesothelial cells. Environ Health Perspect 2008; 116: 1211-1217.

[68] Yehia HN, Draper RK, Mikoryak C. Single-walled carbon nanotube interactions with HeLa cells. J Nanobiotechnol 2007; 5: 8-25.

[69] Zhang Y, Schauer JJ, Shafer MM, Hannigan MP, Dutton SJ. Source apportionment of *in vitro* reactive oxygen species bioassay activity from atmospheric particulate matter. Environ Sci Technol 2008; 42: 7502-7509.

[70] Aarn BB, Fonnum F. Carbon black particles increase reactive oxygen species formation in rat alveolar macrophages *in vitro.* Arch Toxicol 2007; 81: 441-446.

[71] Baulig A, Sourdeval M, Meyer M, Marano F, Baeza-Squiban A. Biological effects of atmospheric particles on human bronchial epithelial cells. Comparison with diesel exhaust particles. Toxicol Vitro 2003; 17: 567-573.

[72] Foldbjerg R, Olesen P, Hougaard M, Dang DA, Hoffmann HJ, Autrup H. PVP-coated silver nanoparticles and silver ions induce reactive oxygen species, apoptosis and necrosis in THP-1 monocytes. Toxicol Lett 2009; 190: 156-162.

[73] Pan Y, Leifert A, Ruau D, *et al.* Gold nanoparticles of diameter 1.4 nm trigger necrosis by oxidative stress and mitochondrial damage. Small 2009; 5: 2067-2076.

[74] Li N, Wang M, Oberley TD, Sempf JM, Nel AE. Comparison of the pro-oxidative and proinflammatory effects of organic diesel exhaust particle chemicals in bronchial epithelial cells and macrophages. J Immunol 2002;169: 4531-4541.

[75] Whitekus MJ, Li N, Zhang M, *et al.* Thiol antioxidants inhibit the adjuvant effects of aerosolized diesel exhaust particles in a murine model for ovalbumin sensitization. J Immunol 2002; 168: 2560-2567.

[76] Li N, Sioutas C, Cho A, *et al.* Ultrafine particulate pollutants induce oxidative stress and mitochondrial damage. Environ Health Perspect 2003; 111: 455-460.

[77] Kagan VE, Tyurina YY, Tyurin VA, *et al.* Direct and indirect effects of single walled carbon nanotubes on RAW 264.7 macrophages: role of iron. Toxicol Lett 2006; 165: 88-100.

[78] Zimmerman MC, Zucker IH. Mitochondrial dysfunction and mitochondrial-produced reactive oxygen species. Hypertension 2009; 53: 112-114.

[79] Karlsson HL, Holgersson A, Moller L. Mechanisms related to the genotoxicity of particles in the subway and from other sources. Chem Res Toxicol 2008; 21: 726-731.

[80] Sayes CM, Gobin AM, Ausman KD, Mendez J, West JL, Colvin VL. Nano-C_{60} cytotoxicity is due to lipid peroxidation. Biomaterials 2005; 26: 7587-7595.

[81] Wang J, Rahman MF, Duhart HM, *et al.* Expression changes of dopaminergic system-related genes in PC12 cells induced by manganese, silver, or copper nanoparticles. Neurotoxicology 2009; 30: 926-933.

[82] WinShwe TT, Mitsushima D, Yamamoto S, *et al.* Extracellular glutamate level and NMDA receptor subunit expression in mouse olfactory bulb following nanoparticle-rich diesel exhaust exposure. Inhal Toxicol 2009; 21: 828-836.

[83] De Rosa M, Zarrilli S, Paesano L, *et al.* Traffic pollutants affect fertility in man. Hum Reprod 2003; 18: 1055-1061.

[84] Li C, Taneda S, Watanabe G, *et al.* Effects of inhaled nanoparticle-rich diesel exhaust on regulation of testicular function in male rats. Inhal Toxicol 2009; 21: 803-811.

[85] Ramdhan DH, Ito Y, Yanagiba Y, *et al.* Nanoparticle-rich diesel exhaust may disrupt testosterone biosynthesis and metabolism *via* growth hormone. Toxicol Lett 2009; 191: 103-108.

[86] Gardea-Torresdey JL, Peralta-Videa JR, Parsons JG, Mokgalaka N.S, de la Rosa G. Production of metal nanoparticles by plants and plant-derived materials. In: Corain B, Schmid G, Toshima N, Eds. Metal nanoclusters in catalysis and materials sciences: the issue of size-control. Amsterdam, The Netherlands: Elsevier 2008; pp. 401-412.

[87]   Parsons JG, Peralta-Videa JR, Gardea-Torresdey JL. Use of plants in biotechnology: synthesis of metal nanoparticles by inactivated plant tissues, plant extracts, and living plants In: Sarkar D, Datta R, Hannigan R, Eds. Concepts and Applications in Environmental Geochemistry, Vol. 5. Amsterdam, The Netherlands: Elsevier 2007: pp. 463-486.

[88]   Parsons JG, Peralta-Videa JR, Dokken KM, Gardea-Torresdey JL. Biological and biomaterials-assisted synthesis of precious metal nanoparticles. In: Kumar CSSR, Ed. Metallic and Metal oxide nanomaterials, Vol. 1. Weinheim: Wiley-VCH Verlag 2009; pp. 461-491.

[89]   Gardea-Torresdey JL, Tiemann KJ, Gamez G, Dokken K, Tehuacanero S, Yacamán MJ. Gold nanoparticles obtained by bio-precipitation from gold(III) solutions. J Nanopart Res 1999; 1: 397-404.

[90]   Armendariz V, Herrera I, Peralta-Videa JR, *et al.* Size controlled gold nanoparticles formation by *Avena sativa* biomasss: use of plants in nanobiotechnology. J Nanopart Res 2004; 6: 377-282.

[91]   Gardea-Torresdey JL, Parsons JG, Gomez E, *et al.* Formation and growth of Au nanoparticles inside live alfalfa plants. Nano Lett 2002; 2: 397-401.

[92]   Harris AT, Bali R. On the formation and extent of uptake of silver nanoparticles by live plants. J Nanopart Res 2007; 10: 691-695.

[93]   Yang X, Li Q, Wang H, *et al.* Green synthesis of palladium nanoparticles using broth of *Cinnamomum camphora* leaf. J Nanopart Res 2010; 12: 1589-1598.

[94]   Manceau A, Nagy KL, Marcus MA, *et al.* Formation of metallic copper nanoparticles at the soil-root interface. Environ Sci Technol 2008; 42: 1766-1772.

[95]   Chan JM, Zhang L, Tong R, *et al.* Spatiotemporal controlled delivery of nanoparticles to injured vasculature. P Natl Acad Sci USA 2010; 107: 2213-2218.

[96]   Lange S. Roadmaps in nanomedicine towards 2020. [Cited: 11th March 2010]. Available from: http://www.nanoforum.org

[97]   Liu Y, Laks P, Heiden P. Controlled release of biocides in solid wood. I. Efficacy against and wood decay fungus (*Gloeophyllum trabeum*). J Appl Polym Sci 2002; 86: 596-607.

[98]   Liu Y, Laks P, Heiden P. Controlled release of biocides in solid wood. II. Efficacy against *Trametes versicolor* and *Gloeophyllum trabeum* wood decay fungi. J Appl Polym Sci 2002; 86: 608-614.

[99]   Liu Y, Laks P, Heiden P. Controlled release of biocides in solid wood. III. Preparation and characterization of surfactant-free nanoparticles. J Appl Polym Sci 2002; 86: 615-621.

[100]   Lin D, Xing B. Root uptake and phytotoxicity of ZnO nanoparticles. Environ Sci Technol 2008; 42: 5580-5585.

[101]   Liu Q, Chen B, Wang Q, *et al.* Carbon nanotubes as molecular transporters for walled plant cells. Nano Lett 2009; 9: 1007-1010.

[102]   Racuciu M, Creanga D. TMA-OH coated magnetic nanoparticles internalized in vegetal tissue. Rom J Phys 2007; 52: 595-402.

[103]   Gonzáles-Melendi P, Fernández-Pacheco R, Coronado MJ, *et al.* Nanoparticles as smart-delivery systems in plants: assessment of different techniques of microscopy for their visualization in plant tissues. Ann Bot 2008; 101: 187-195.

[104]   Corredor E, Testillano PS, Coronado MJ, *et al.* Nanoparticle penetration and transport in living plants: *in situ* subcellular identification. BMC Plant Biol 2009; 9: 45-57.

[105]   Zhu H, Han J. Xiao JQ, Jin Y. Uptake, translocation, and accumulation of manufactured iron oxide nanoparticles by pumpkin plants. J Environ Monitor 2008; 10: 713-717.

[106]   Battke F, Leopold K, Maier M, Schmidhalter U, Schuster M. Palladium exposure of barley: uptake and effects. Plant Biol 2008; 10: 272-276.

[107]   Leopold K, Schuster M. Pd Particles as Standardized Test Material for Bioavailability Studies of Traffic Related Pd Emissions to Barley Plants. In: Wiseman C, Zereini F, Eds. Urban Airborne Particulate Matter: Origins, Chemistry, Fate and Health Impacts? 1st ed. Berlin: Springer Verlag 2010, pp. 399-410.

[108]   Speranza A, Leopold K, Maier M, Taddei AR, Scoccianti V. Pd-nanoparticles cause increased toxicity to kiwifruit pollen compared to soluble Pd(II). Environ Pollut 2010; 158: 873-882.

[109]   Hischemöller A, Nordmann J, Ptacek P, Mummenhoff K, Haase M. *In vivo* imaging of the uptake of upconversion nanoparticles by plant roots. J Biomed Nanotechnol 2009; 5: 278-284.

[110]   Park KH, Chhowalla M, Iqbal Z, Sesti F. Single-walled carbon nanotubes are a new class of ion channel blockers. J Biol Chem 2003; 12: 50212-50216.

[111]   Pulskamp K, Wörle-Knirsch JM, Hennrich F, Kern K, Krug HF. Human lung epithelial cells show biphasic oxidative burst after single-walled carbon nanotube contact. Carbon 2007; 45: 2241-2249.

[112]   Ruffini Castiglione M, Cremonini R. Nanoparticles and higher plants. Caryologia 2009; 62: 161-165.

[113] Moor, MN. Do nanoparticles present ecotoxicological risks for the health of the aquatic environment? Environ Int 2006; 32: 967-976.

[114] Moore MN, Readman AJ, Readman JW, Lowe DM, Frickers PE, Beesley A. Lysosomal toxicity of carbon nanoparticles in cells of the molluscan immune system: an *in vitro* study. Nanotoxicology 2009; 3: 40-45.

[115] Baun A, Sørensen SN, Rasmussen RF, Hartmenn NB, Koch CB. Toxicity and bioaccumulation of xenobiotic organic compounds in the presence of aqueous suspensions of aggregates of nano-C_{60}. Aquat Toxicol 2008; 86: 379-387.

[116] Hund-Rinke K, Simon M. Ecotoxic effect of photocatalytic active nanoparticles (TiO_2) on algae and daphnids. Environ. Sci Pollut Res 2006; 13: 225-232.

[117] Van Hoecke K, De Schamphelaere KAC, Van der Meeren P, Lcucas S, Janssen CR. Ecotoxicity of silica nanoparticles to the green alga *Pseudokirchneriella subcapitata*: importance of surface area. Environ Toxicol Chem 2009; 27: 1948-1957.

[118] Velzeboer I, Hendrix AJ, Ragas AM, van de Meent D. Nanomaterials in the environment: aquatic ecotoxicity tests of some materials. Environ Toxicol Chem 2009; 27: 1942-1947.

[119] Hall S, Bradley T, Moore JT, Kuykindall T, Minella L. Acute and chronic toxicity to nano-scale TiO_2 particles to freshwater fish, cladocerans, and green algae, and effects of organic and inorganic substrate on TiO_2 toxicity. Nanotoxicology 2009; 3: 91-97.

[120] Adams LK, Lyon D, Alvarez PJJ. Comparative ecotoxicity of nanoscale TiO_2, SiO_2, and ZnO in water suspensions. Water Res 2006; 40: 3527-3532.

[121] Bouldin JL, Ingle TM, Sengupta A, Alexander R, Hannigan RE, Buchanan RA. Aquatic toxicity and food chain transfer of quantum dots in freshwater algae and *Ceriodaphnia dubia*. Environ Toxicol Chem 2008; 27: 1958-1963.

[122] Porter AE, Gass M, Muller K, Skepper JN, Midgley P, Welland M. Visualizing the uptake of C_{60} to the cytoplasm and nucleus of human monocyte-derived macrophage cells using energy-filtered transmission electron microscopy and electron tomography. Environ Sci Technol 2007; 47: 3012-3017.

[123] Miao AJ, Schwehr KA, Xu C, *et al.* The algal toxicity of silver engineered nanoparticles and detoxification by exopolymeric substances. Environ Pollut 2009; 157: 3034-3041.

[124] Verdugo P, Alidredge AL, Azam F, Kirchman DL, Passow U, Santschi PH The oceanic gel phase: a bridge in the DOM-POM continuum. Mar Chem 2004; 92: 67-85.

[125] Pistocchi R, Mormile MA, Guerrini F, Isani, G, Boni L. Increased production of extra- and intracellular metal-ligands in phytoplankton exposed to copper and cadmium. J Appl Phycol 2000; 12: 469-477.

[126] Navarro E, Piccapietra F, Wagner B, *et al.* Toxicity of silver nanoparticles to *Chamydomonas reinhardtii*. Environ Sci Technol 2008; 42: 8959-8964.

[127] Morones JR, Elechiguerra JL, Camacho A, *et al.* The bactericidal effect of silver nanoparticles. Nanotechnology 2005; 16: 2346-2353.

[128] Franklin NM, Rogers NJ, Apte SC, Batley GE, Gadd GE, Casey PS. Comparative toxicity of nanoparticulate ZnO, bulk ZnO, and $ZnCl_2$ to a freshwater microalga (*Pseudokirchneriella subcapitata*): the importance of particle solubility. Environ Sci Technol 2007; 41: 8484-8490.

[129] Derfus AM, Chan WC, Bhatia SN. Probing the cytotoxicity of semiconductor quantum dots. Nano Lett 2004; 4: 11-18.

[130] Tiju J, Morrison M. Nanotechnology in agriculture and food, 2006 [cited: 11th March 2010]. Available from: http://www.nanoforum.org

[131] Lu CM, Zhang CY, Wen JQ, Wu GR, Tao MX. Research of the effect of nanometer materials on germination and growth enhancement of *Glycine max* and its mechanisms. Soybean Sci 2002; 21: 168-172.

[132] Hong FS, Zhou J, Liu C, *et al.* Effect of nano-TiO_2 on photochemical reaction of chloroplasts of spinach. Biol Trace Elem Res 2005; 105: 269-279.

[133] Zheng L, Hong FS, Lu SP, Liu C. Effect of nano-TiO_2 on strength of naturally aged seeds and growth of spinach. Biol Trace Elem Res 2005; 104: 83-91.

[134] Yang F, Hong FS, You WJ, *et al.* Influences of nano-TiO_2 on the nitrogen metabolism of growing spinach. Biol Trace Elem Res 2006; 110: 179-190.

[135] Yang L, Watts DJ. Particle surface characteristics may play important role in phytotoxicity of alumina nanoparticles. Toxicol Lett 2005, 158: 122-132.

[136] Lin D, Xing B. Phytotoxicity of nanoparticles: inhibition of seed germination and root growth. Environ Pollut 2007; 150: 243-250.

[137] Cañas JE, Long M, Nations S, *et al.* Effects of functionalized and non functionalized single-walled carbon nanotubes on root elongation of selected crop species. Environ Toxicol Chem 2008; 27: 1922-1931.

[138] Lin D, Xing B. Root uptake and phytotoxicity of ZnO nanoparticles. Environ Technol 2008; 42: 5580-5585.

[139] Lee WM, An YJ, Yoon H, Kweon HS. Toxicity and bioavailability of copper nanoparticles to the terrestrial plants mung bean (*Phaseolus radiatus*) and wheat (*Triticum aestivum*): plant agar test for water-insoluble nanoparticles. Environ Toxicol Chem 2008; 27: 1915-1921.

[140] Ma Y, Kuang L, He X, *et al.* Effects of rare earth oxide nanoparticles on root elongation of plants. Chemosphere 2010; 78: 273-279.

[141] Racuciu M, Creanga D. Cytogenetically changes induced by aqueous ferrofluids in agricultural plants. In: Hafeli U, Schutt W, Zborowski M, Safarikova M, Safarik I, Eds. Scientific and Clinical Applications of Magnetic Carriers. New York: Plenum Press 2005; p. 214.

[142] Kumari M, Mukherjee A, Chandrasekaran N. Genotoxicity of silver nanoparticles in *Allium cepa*. Sci Total Environ 2009; 407: 5243-5246.

[143] Asli S, Neumann PM. Colloidal suspension of clay or titanium dioxide nanoparticles can inhibit leaf growth and transpiration *via* physical effects on root water transport. Plant Cell Environ 2009; 32: 577-584.

[144] Schäfer J, Hannker D, Eckhart JD, Stüben D. Uptake of traffic-related heavy metals and platinum group elements (PGE) by plants. Sci Total Environ 1998; 215: 59-67.

[145] Leopold K, Maier M, Schuster M. Preparation and characterization of Pd/Al$_2$O$_3$ and Pd nanoparticles as standardized test material for chemical and biochemical studies of traffic related emissions. Sci Total Environ 2008; 394: 177-182.

[146] Shah V, Belozerova I. Influence of metal nanoparticles on the microbial soil community and germination of lettuce seeds. Water Air Soil Pollut 2009; 97: 143-148.

[147] Kristen U, Jung K, Pape W, Pfannenbecker U, Rensch A, Schell R. Performance of the pollen tube growth test in the COLIPA validation study on alternatives to the rabbit eye irritation test. Toxicol. Vitro 1999; 13: 335-342.

[148] Barile FA, Dierickx PJ, Kristen U. *In vitro* toxicity testing for prediction of acute human toxicity. Cell Biol Toxicol 1994; 10: 155-162.

[149] Bergweiler CJ, Manning WJ. Inhibition of flowering and reproductive success in spreading dogbane (*Apocynum androsaemifolium*) by exposure to ambient ozone. Environ Pollut 1999; 105: 333-339.

[150] Calzoni GL, Antognoni F, Pari E, Fonti P, Gnes A, Speranza A. Active biomonitoring of heavy metal pollution using *Rosa rugosa* plants. Environ Pollut 2007; 149: 239-245.

[151] Speranza A, Crosti P, Malerba M, Stocchi O, Scoccianti V. The environmental endocrine disruptor, bisphenol A, affects germination, elicits stress response, and alters steroid hormone production in kiwifruit pollen. Plant Biol 2011; 13: 209-217

[152] Stoimenov PK, Klinger RL, Marchin GL, Klabunde KJ. Metal oxide nanoparticles as bactericidal agents. Langmuir 2002; 18: 6679-6686.

[153] Zhang D, Neumann O, Wang H, *et al.* Gold nanoparticles can induce the formation of protein-based aggregates at physiological pH. Nano Lett 2009; 9: 666-671.

[154] Fumagalli A, Faggion B, Ronchini M, Terzaghi G, Lanfranchi M, Cherchi L. Platinum, palladium, and rhodium deposition to the *Prunus laurus cerasus* leaf surface as an indicator of the vehicular traffic pollution in the city of Varese area. Environ Sci Pollut Res 2010; 17: 665-673.

Plants as Indicators of Nanoparticles Toxicity

CHAPTER 2

Plants as Indicators of Nanoparticles Toxicity

Mamta Kumari, Vinita Ernest, Amitava Mukherjee and N.Chandrasekaran[*]

Center for Nanobiotechnology, VIT University, Vellore 632 014, India

Abstract: Increasing application of nanoparticles in consumer products enhances its release into the environment. Plants are the primary target species to work out a comprehensive toxicity profile for nanoparticles. Toxicity profiles of nanoparticles to the plant system, uptake and its subsequent fate within the food chain are not available. The phytoxicological behaviour of silver and zinc oxide nanoparticles on *Allium cepa* and seeds of *Lycopersicum esculentum* (tomato), *Cucumis sativus* (cucumber) and *Zea mays* (maize) were experimented. The *in vitro* studies of *Allium cepa* root tips exposed to a concentration-tested range of 25, 50, 75, and 100 µg ml^{-1} nanoparticles for 4 h revealed different cytotoxicological effects including mitotic index, chromosomal aberrations, vagrant chromosomes, sticky chromosomes, disturbed metaphase, breaks, and formation of micronucleus. Nanoparticles treated seeds showed reduced germination rate and decrease in shoot and root lengths. Nanoparticles treated seedlings showed reduced shoot and root lengths. The percentage germination of seeds was delayed with increasing concentration of nanoparticles. Though engineered nanoparticles have significant advantage in biomedical applications, it also requires a great deal of toxicity profile on the other side to ascertain the biosafety and risk of using nanoparticles in consumer products.

Keywords: Nanoparticles, Silver, Zinc oxide, *Allium cepa*, Seeds, Phyto-toxicity, Cyto-toxicity, Surface characteristics, Accumulation, Adsorption, FT-IR, Mitotic index, Relative germination rate, Chromosomal aberrations.

INTRODUCTION

There is a rapid development in the field of nanotechnology and it has resulted in a vast array of nanoparticles with varying size, shape, surface charge chemistry, coating and solubility behaviour. Nanoparticles are defined as particles less than 100 nm in one dimension at atomic, molecular and macromolecular scales [1, 2]. The nanoparticle differs from its own bulk-form in its physical properties [3, 4] and could be more toxic than its bulk form [5, 6]. Nanotechnology has wide applications in various industries thereby enhancing the economy of a country. On the other hand, it also creates negative impacts on human and non-human biota [7]. There are nearly 800 consumer products where nanoparticles are being used [8]. The antimicrobial properties of silver nanoparticles are being increasingly exploited in consumer products like deodorants, clothing materials, bandages, and also in cleaning solutions and sprays [9, 10].

USAGE OF NANOPARTICLES

Till date, nanoparticles are used in more than 1015 commercial products [11] such as:

- Consumer products: sunscreen, cosmetics, textiles, toys, sport and ICT equipments.

- Health care: medicines, oral vaccines, drug delivery biocompatible materials.

- Energy conversion: economic lighting batteries, solar and fuel cells.

- Construction materials: improved rigidity and insulating properties.

- Automobile/aerospace industry: fuel additives.

[*]**Address correspondence to N. Chandrasekaran:** Nanobiomedicine Lab, School of Biosciences and Technology, VIT University, Vellore 632 014, India; Tel: 0091-416-2202624; Fax: 0091-416-224-3092; E-mail: nchandra40@hotmail.com

Haseeb Ahmad Khan and Ibrahim Abdulwahid Arif (Eds)

Samsung's "Nano Silver" washing machine releases nano silver directly into waste water systems. The effluent containing nano silver would kill beneficial bacteria and disrupt ecosystem functioning [12].

FLOW OF NANOPARTICLES IN THE ENVIRONMENT

Plants as an important component of the environmental and ecological system need to be included when evaluating the overall fate, transport, and exposure pathways of nanoparticles in the environment. Thus, before dumping a huge amount of hazardous nanomaterials into the environment, there is a need to investigate the solubility and degradability of engineered nanoparticles in soils and waters and to establish baseline information on their safety, toxicity and adaptation towards soil and aquatic life. The fate of nanoparticles in the ecosystem which consists of soil, water and air is depicted in Fig. **1**.

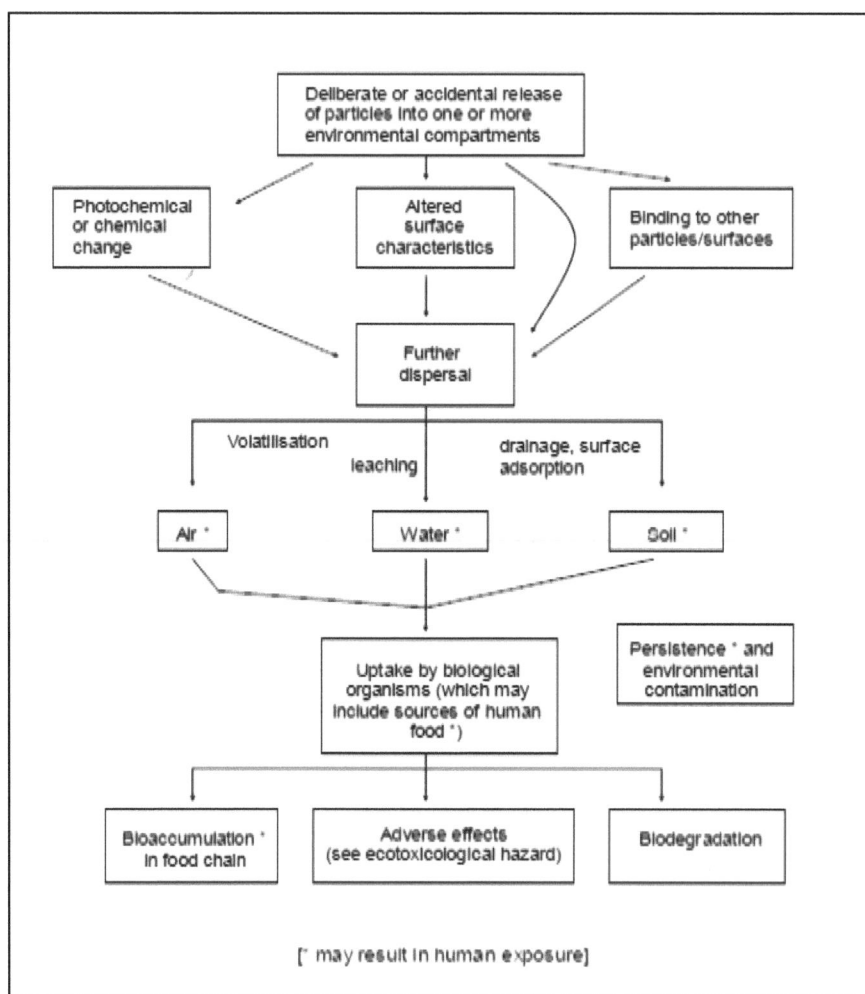

Figure 1: Fate of nanoparticles in ecosystem. [Source: European Commission Report, 2005]

The increased usage of nanoparticles in many consumer products leads to their release into the environmental components [13]. The toxicity of zinc oxide and cerium oxide nanoparticles is based on dissolution and oxidative stress properties [14]. Though the exact mechanism behind nanoparticles toxicity is yet to be elucidated, many studies have suggested that oxidative stress caused by reactive oxygen species (ROS) and lipid peroxidation (LPO) plays an important role in nanoparticle toxicity [15]. Another mechanism is, through the release of metal ions from nanoparticles [16]. The toxicity data of nanoparticles to ecological and terrestrial species are limited [17]. There are still many unresolved issues and challenges concerning the biosafety and the biological effects of nanoparticles to the plant system.

Plants are an important component in the ecological system where it can serve as a potential pathway for nanoparticles transport and route for bioaccumulation into the food chain. Plants need to be included when evaluating the overall fate, transport and exposure pathways of nanoparticles in the environment [18]. To study the toxicity of nanomaterials, plants are the suitable indicator organisms. Plant systems have well defined end points like phytotoxicity testing, seed germination toxicity test, root length, shoot length, biochemical test *etc.* Plants are also recognized as excellent genetic models to detect environmental mutagens and are frequently used in bio-monitoring studies [19].

SELECTION OF TEST SYSTEMS

Test systems were selected based on OECD (Organization for Economic Cooperation and Development) and USEPA (Environment Protection Act protocol 1996) regulations.

Plant: *Allium cepa*

Seeds: *Lycopersicum esculentum*, *Cucumis sativus* and *Zea mays*

The following reasons were the major attributes for selecting the above mentioned test systems.

- Best bio-indicator to check the environmental pollution and toxicant

- Easy availability

- Low chromosome number

- Sensitive to phytotoxicity

- Low germination time

- High germination rate

ALLIUM CEPA - A POTENTIAL BIO-INDICATOR

Allium cepa is a potential test species for studying cytotoxicity and genotoxicity of nanoparticles. *Allium cepa* is an efficient species to study chromosomal aberrations and cytotoxicity testing [20] and has been used routinely for studying the effects of toxic materials in environmental monitoring program [21]. Some studies have already reported both the positive and negative aspects of nanoparticles on higher plants. Nanoscale SiO_2 and TiO_2 enhanced nitrate reductase activity in soybean and apparently hastened its germination and growth [22]. Nano-TiO_2 promoted photosynthesis and nitrogen metabolism and improved growth of spinach [23-25]. Silver nanoparticles were found to be genotoxic to *A. cepa* root tip cells [26].

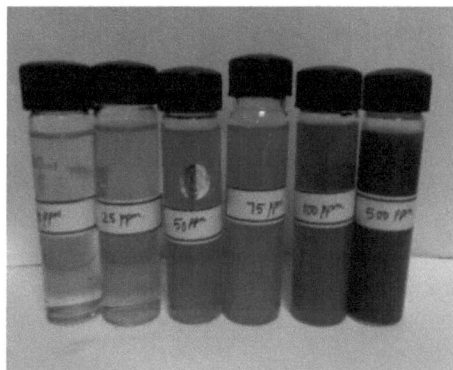

Figure 2: Dispersion of silver nanoparticles.

NANOPARTICLES DISPERSION

The engineered silver nanoparticles and nano zinc oxide nanoparticles were dispersed in deionized (Milli-Q) water and sonicated using ultrasonic vibrations (Sonics Vibracell ultrasonicator, 130W, 20 kHz) for 30 min to produce different concentrations of 10, 25, 50, 75, 100 and 500 $\mu g\ ml^{-1}$ nanoparticles dispersion (Figs. **2** and **3**). *Note:* The stability of nanoparticles in aqueous dispersion is an important factor in studying its effects to the test system. When nanoparticles concentration is high in the aqueous dispersion, the possibility of it getting agglomerated is high, causing least effect to the test system.

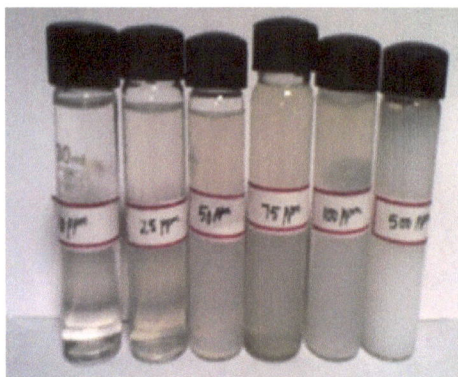

Figure 3: Dispersion of nano zinc oxide.

PHYSICOCHEMICAL CHARACTERIZATION OF NANOPARTICLES

The characteristic peaks of the silver and zinc oxide nanoparticles were identified using UV-visible double beam spectrophotometer (Systronics 2201). The peaks were observed at 424 and 374nm for silver and zinc oxide respectively. The morphological features of the nanoparticles were characterized using transmission electron microscope (Technai10, Philips). Silver nanoparticles showed spherical to oval shape while zinc oxide nanoparticles showed spherical to hexagonal shaped particles. Scanning electron microscopy (Hitachi, S-3400) and atomic force microscopy were also done to find the surface morphology and shape of the nanoparticles. Particle size distribution, effective diameter and polydispersity were assessed by 90Plus particle size analyzer (Brookhaven Instruments Corporation). Fourier transform infrared spectroscopy (FT-IR) (Thermonicolar, Avatar-330, USA) was used for surface characterization and to show the presence of functional groups like carboxylic ester group, amide stretching bands, and strong fingerprint region for nanoparticles in control and test. The crystalline structure of the nanoparticles was examined by X-ray diffraction (XRD) analysis using JEOL JDX 8030 spectrometer (Fig. **4**).

AVAILABILITY OF NANOPARTICLES IN DISPERSION

When nanoparticles are dispersed in aqueous media, it is important to check their availability in suspension. *i.e.,* the real concentration that is available for the test system (*e.g., Allium cepa* root tip test). After dispersion, there is a possibility of agglomeration and settling down of particles. Therefore, the real concentration that was initially taken during the experiment may not be correctly reflected. Thus, bioavailable concentration is taken into account. Another reason is that when nanoparticles (silver and zinc oxide) are dispersed in water, the chances of ions getting released from the nanoparticles into the medium are more likely.

The concentration was measured as total metals using an atomic absorption spectrophotometer (Varian, AA-240) after acidification with 1% nitric acid. The nanoparticles dispersion was centrifuged at 12, 000 rpm for 10 minutes and filtered through 0.22μm Anapore membrane disc and the clear filtered supernatant was carefully collected in a boiling tube and 2 ml of 1% nitric acid were added and analyzed using atomic absorption spectrophotometer (Table **1**) [27, 28].

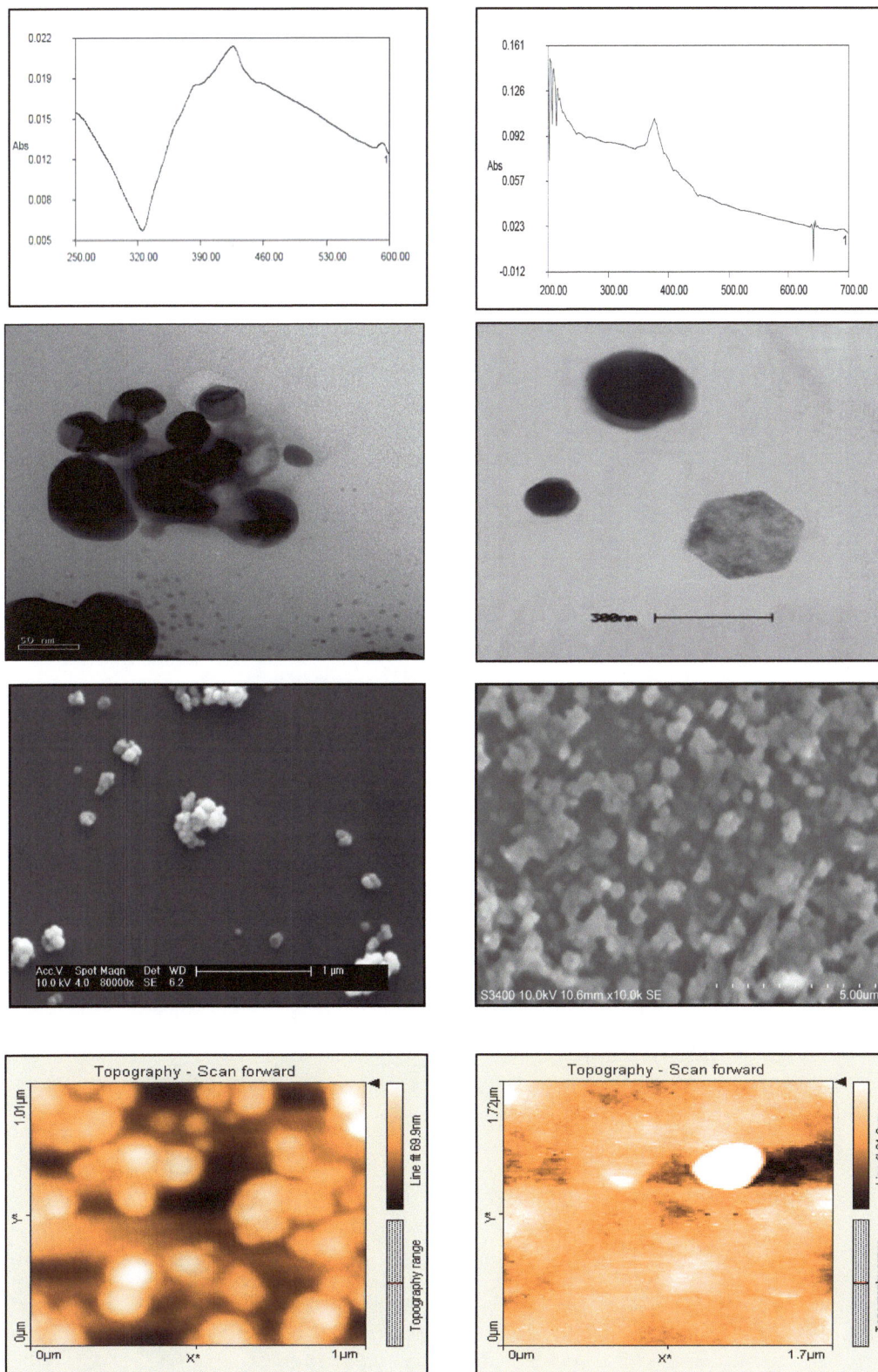

Figure 4: Characterization of nanoparticles by different techniques. (a) UV-Vis spectrum for silver nanoparticles, (b) UV-Vis spectrum for zinc oxide nanoparticles, (c) TEM image of silver nanoparticles showing spherical to oval in shape, (d) TEM image of zinc oxide nanoparticles showing spherical to hexagonal shape, (e) SEM image of silver nanoparticles, (f) SEM image of nano zinc oxide particles, (g) AFM image of silver nanoparticles, (h) AFM image of nano zinc oxide particles.

Table 1: Metal ion concentration measurement in dispersion (n=3).

Concentration taken for experiment ($\mu g\ ml^{-1}$)	Bioavailable concentration (Mean ± SE)	
	Silver nanoparticles	Zinc oxide nanoparticles
10	4.04 ± 0.95	4.04 ± 0.95
25	9.5 ± 1.15	9.5 ± 1.15
50	22.2 ± 1.6	20.7 ± 1.6
75	27.9 ± 0.91	31.9 ± 0.91
100	41 ± 1.03	43.7 ± 1.03
500	168.7 ± 0.98	159.7 ± 0.98

PHYTOTOXICITY ASSESSMENT OF NANOPARTICLES

In vitro studies on *Allium cepa* (Root test assay)

Allium cepa was used for this study, as it is the suitable bio-indicator for testing toxicity of materials. It is also easy to analyze the cellular and chromosomal deformation caused due to novel materials because of its low chromosomal number (2n=16). Four healthy onion bulbs (20-25 g) were grown in dark in a cylindrical glass beaker at room temperature (28±0.5°C) and renewed water supply every 24 h. When the roots reached 2 to 3 cm in length, they were treated with different concentrations of nanoparticles suspension for 4 h [21, 29]. Five replicates were used for each concentration Fig. **5**.

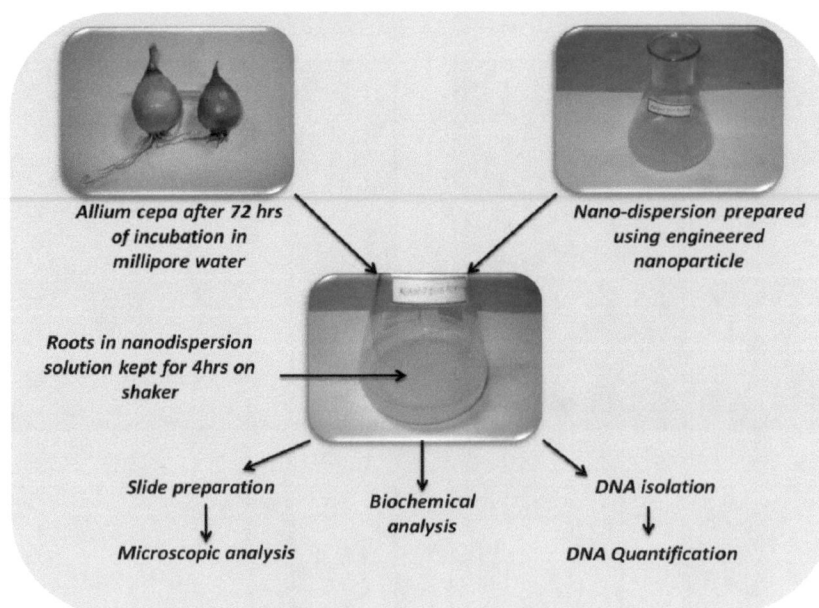

Figure 5: Root cells of *Allium cepa* treated with nanoparticle dispersion.

MICROSCOPIC EXAMINATION

Five bulbs and eight new root tips were used for each concentration of SNP and nano zinc oxide particles in dispersion. The slides were prepared for each concentration and control following Saffranin squash technique. The root tips were kept in 1M HCl for 6 mins followed by staining with 40-45% saffranin. Staining was continued for 5-6 min. The slides were analyzed at 1000X magnification for cytological changes. The mitotic index was calculated as the number of dividing cells per number of 1000 observed cells [21]. The number of aberrant cells was noted per total cells scored at each concentration [33]. The MN index was calculated as mentioned below [34].

$$\text{Mitotic index (MI)} = \text{TDC/TC} \times 100 \tag{1}$$

$$\text{Phase index (PI)} = \text{TC/TDC} \times 100 \tag{2}$$

$$\text{Total percentage of abnormal cells} = T_{abn}/\text{TDC} \times 100 \tag{3}$$

$$\text{MN Index (\%)} = T_{MN}/\ T_{BN} \times 100 \tag{4}$$

Where TDC is total number of dividing cells, TC is total number of cells observed, T_{abn} is total number of abnormal cells, T_{MN} is total number of micronucleus observed, and T_{BN} is total number of binucleated cells observed.

EFFECT OF NANOPARTICLES ON *A. CEPA*

The microscopic results of the cytological and chromosomal aberrations observed in the root tip cells of *A. cepa* treated with different concentrations of silver nanoparticles and nano zinc oxide particles are shown in Figs. **6** and **7**. Cell cycle analysis showed that the percentage of cells in the different phases of mitosis (prophase, metaphase, anaphase and telophase) decreased with increasing silver nanoparticles. Prophase percentage changed rapidly and significantly, with a similar result as seen for the mitotic indices. The mitotic index (MI) for the samples treated with 25, 50, 75, and 100 µg ml^{-1} silver nanoparticles were 60.6, 43.76, 38.06, 35.01, 27.06 % respectively. In the case of control, it was 60.3 %. No chromosomal aberration was observed in the control.

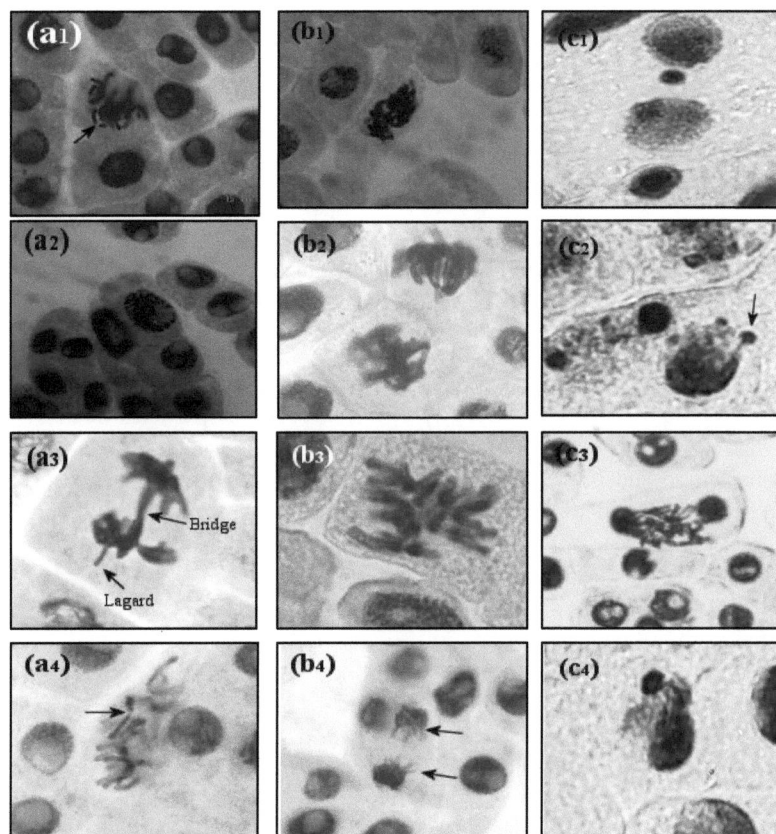

Figure 6: Chromosomal aberration observed in *A. cepa* meristematic cells exposed to silver nanoparticles. (a$_1$) Break in metaphase stage, (a$_2$) Digested cells, (a$_3$) Bridge at anaphase stage, (a$_4$) break at anaphase stage, (b$_1$) sticky chromosomes at metaphase stage, (b$_2$) sticky chromosomes at anaphase, (b$_3$) clump formation at metaphase, (b$_4$) sticky chromosomes at anaphase with lagards, (c$_1$) micronucleus at telophase, (c$_2$) micronucleus at interphase and telophase, (c$_3$) micronucleus at metaphase and (c$_4$) micronucleus at prophase stage.

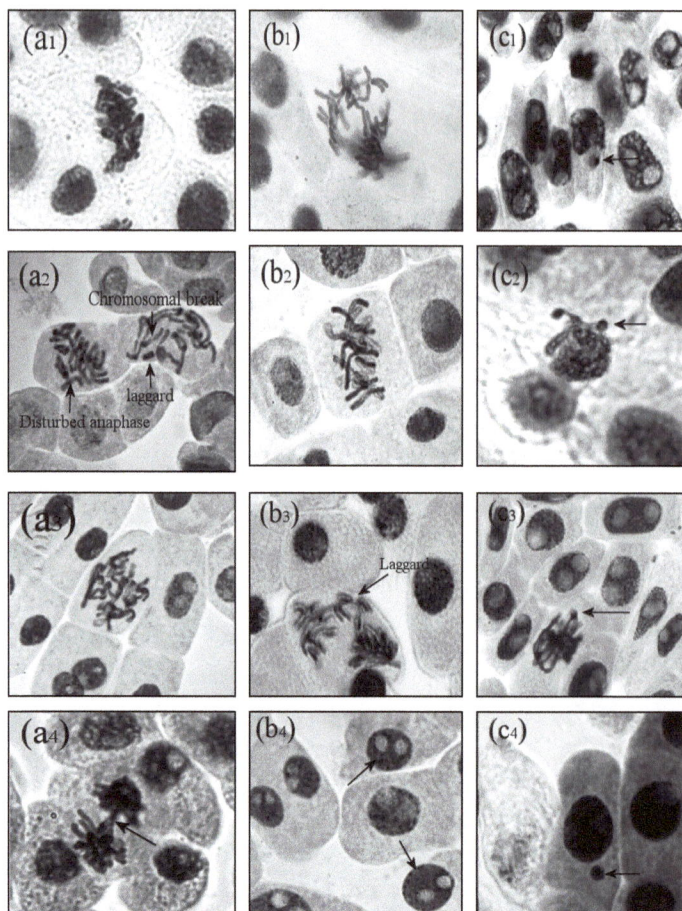

Figure 7: Chromosomal aberrations observed in *A. cepa* meristematic cells exposed to nano zinc oxide particles. (a$_1$) Sticky chromosomes in metaphase stage, (a$_2$) Disturbed anaphase stage with chromosomal break and laggard, (a$_3$) disturbed metaphase, (a$_4$) sticky chromosomes at anaphase stage with bridge, (b$_1$) Multipolar anaphase, (b$_2$) C-mitotic cell, (b$_3$) vagrant chromosomes with laggard at anaphase, (b$_4$) binucleated cells at early prophase stage, (c$_1$andc$_4$) Prophase nuclei with micronucleus in interphase, (c$_2$) cell in early anaphase with chromosome adherence and budding micronucleus, (c$_3$) binucleated cell in early telophase with micronucleus in metaphase.

The micronucleus (MN) index for 50, 75, and 100 μg ml^{-1} silver nanoparticles treated *A. cepa* root cells showed 9.8, 11.2 and 17.7%. The effect of silver nanoparticles concentration on MI was significantly different (p<0.05) for 50, 75 and 100 μg ml^{-1} when compared to the control. At 50 μg ml^{-1} concentration, chromatin bridge, stickiness, and disturbed metaphase were observed; for 75 μg ml^{-1} vagrant chromosome, chromosomal break and at 100 μg ml^{-1}, binucleated cells, complete disintegration of cell walls, nuclear membrane disruption, bridge and at anaphase, vagrant chromosomes were observed.

The MI for the samples treated with 25, 50, 75, and 100 μg ml^{-1} nano zinc oxide particles were 50.4, 40.1, 35.3, 29.2 % respectively. The chromosomal aberration index for 25, 50, 75, and 100 μg ml^{-1} nano zinc oxide treated *A. cepa* root cells showed 0.52, 1.52, 2.74, and 4.12 % respectively. The MN index for 25, 50, 75, and 100 μg ml^{-1} nano zinc oxide particles treated *A. cepa* root cells showed 8.7, 9.2, 15.6 and 18.7%. The effect of nano zinc oxide concentration on MI was significantly different (p<0.05) for 50, 75 and 100 μg ml^{-1} as compared to the control. Chromosomal aberration index and MN index were significantly different for all concentrations of nano zinc oxide particles. It has been reported that micronucleus can be an effective parameter to assess the clastogenic and aneugenic effects in *A. cepa* [18].

The changes in the organization and morphology of the chromosomes were observed in the root tips exposed to the nano zinc oxide particles. At varying concentrations of nano zinc oxide particles different chromosomal aberrations were observed. Chromosomal change in stickiness is a common sign of toxic influence on the chromosomes and is probably an irreversible effect. However, stickiness has been shown to occur due to DNA condensation [1].

ELECTRON MICROSCOPY

For SEM and TEM analysis, the samples were prepared as given below. The fragments (1mm long) of root tips were fixed in 2% glutaraldehyde for 2 h and 0.1M cacodylate buffer (pH 7.2) and post-fixed in 1% osmium tetroxide (12 h). After rapid dehydration in alcohol, the material was embedded in Epon 812. Semi-thin and ultrathin sections were prepared from the middle fragment of the central longitudinal and cross section of the root.

For viewing through electron microscope, half of the sections prepared as above were stained with uranyl acetate and Reynolds' reagent, the other half were left unstained [35]. The results are shown in Fig. **8**.

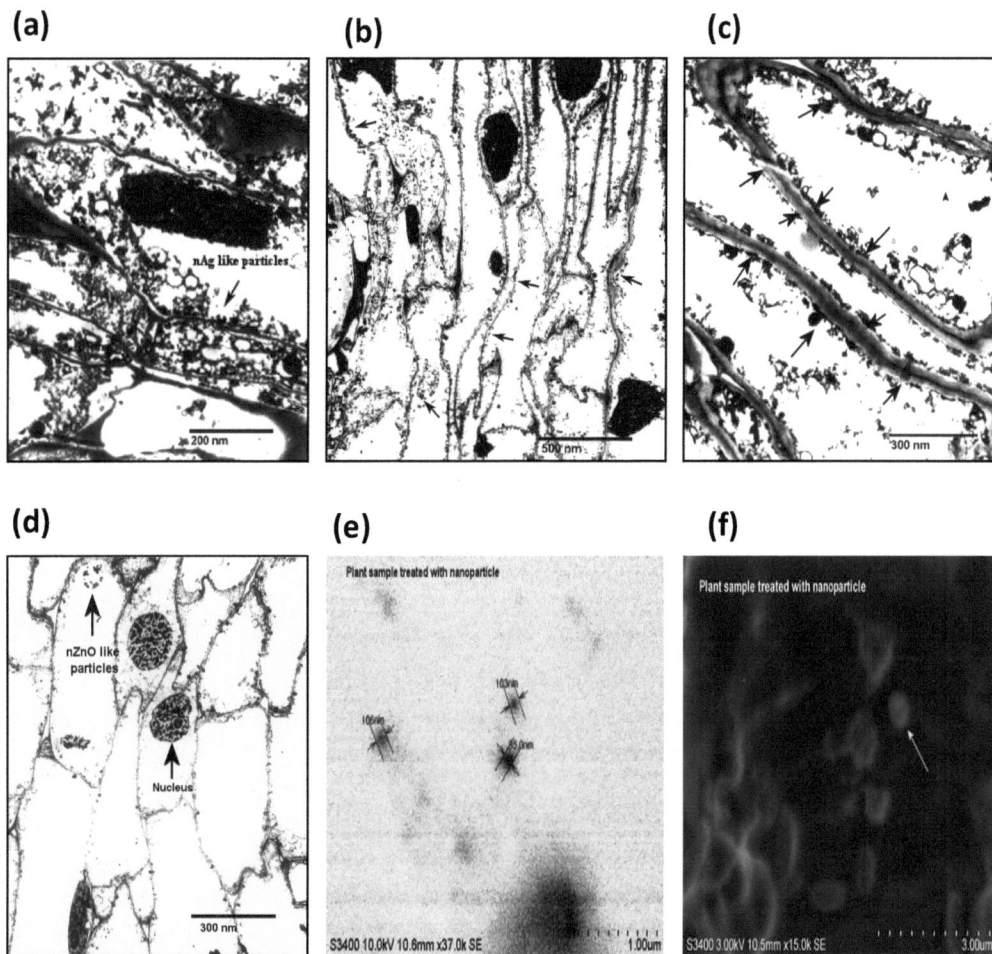

Figure 8: (a) TEM image of silver nanoparticles treated root cells *A. cepa* showed the presence of nanoparticles in to the vacuoles, in the nucleus and around the nucleus (15000X); (b) TEM images of root cells of *C. sativus* treated with silver nanoparticles particles showed presence of nanoparticles in the different compartment of the root cells(1500X); (c) TEM image of nano zinc oxide treated root sample of *A. cepa* showing presence of nanoparticles on the inner and outer edge of cell wall and inside vacuoles (10000X); (d) TEM image of *A. cepa* root cells treated with nano zinc oxide particles showed the presence of nanoparticles in and around the nucleus, in the inner edge of the cell wall (1500X); (e) SEM image of nano zinc oxide particles treated root cells showing particles in the size range of 95, 103 and 106 nm (30000X); (f) SEM image of nano zinc oxide particles treated root cells showing the presence of particles in cellular matrix (30000X).

The TEM and SEM analysis for root cells of *A. cepa* showed particles deposition on cell membranes, around nucleus, agglomeration of particles at the inner side of cell membrane, into the vacuoles, also surrounded by the vacuoles, and also in the cytoplasm. There was a chain of particles like deposits in extracellular matrix as well as in cellular matrix for the 100 μg ml^{-1} silver nanoparticles and nano zinc oxide particles treated cells, which might be agglomerated silver nanoparticles in the cells of the root cells of *A. cepa*, and the root of the seed system like *C. sativus*, *L. esculentum* and *Z. mays*. TEM images (Fig. **8**) of root cells of *C. sativus* treated with silver nanoparticles particles showed presence of nanoparticles in different compartments of the root cells of *C. sativus*. Silver nanoparticles treated root cells showed the agglomeration of nanoparticles around the cell membranes, around and in to the vacuoles, around the inner side of cell membrane, in cytoplasm, and also around and into the nucleus.

Deposits of nano zinc oxide particles were noted in SEM images of *A. cepa* treated with 100 μg ml^{-1} concentration; showing nano zinc oxide particles in the cellular matrix with sizes noted to be 95 nm, 103 nm, and 106 nm. TEM images of *A. cepa* roots treated with 100 μg ml^{-1} nano zinc oxide particles showed presence of nanoparticles in cell membrane, and agglomeration of particles into the nucleus and around the nucleus. The observation parallels the reports [36, 37] that nanoparticles penetrate the plant cells and get deposits in matrix.

IN VITRO STUDIES: SEED EXPERIMENT

Exposure of silver nanoparticles and nano zinc oxide particles to the seeds of *C. sativus*, *Z. mays* and *L. esculentum*.

Seed Germination Test

This test was conducted following the standard method [32]. The test was performed on three seeds (*C. sativus*, *L. esculentum* and *Z. mays*). The relative seed germination rate (RSG) and relative root growth (RRG) were calculated using the equations (5) and (6). Germination index (GI) was determined using equation (7). In addition, 50% effective concentration of nanoparticles (IC$_{50}$) and 95% confidence level were determined by using probit computer program (US-EPA, 1994).

Relative Seed Germination Rate = (Ss/Sc) x 100   (5)

Relative Root Growth = (Rs/Rc) x 100   (6)

Germination Index = (RSG x RRG)/100   (7)

Where Ss is the number of seed germinated in sample, Sc is the number of seed germinated in control, Rs is the average root length in sample, and Rc is the average root length in control.

Seedling Exposure

The seeds were first checked for their viability by suspending them in deionized water. The seeds which settled to the bottom were selected for further study. The seeds were then soaked for 10 min in 10% sodium hypochlorite solution, which acts as a surface sterilizing agent [30]. After surface sterilization, the seeds were rinsed in deionized water thrice and were then stirred for 2 h in nanoparticles dispersion (10 μg ml^{-1} , 100 μg ml^{-1}, 500 μg ml^{-1}) using a magnetic stirrer. Whatman No.1 filter paper was then placed into each Petri dish (100 mm x 15 mm) and 5 ml of the respective particle suspensions were added using a Pasteur pipette. The seeds were then transferred to the Petri dish, with 10 seeds per dish and they were placed equidistant from one another (Fig. **9**). The dishes were covered and sealed with sealing tape and placed in dark condition. The end points of the experiment were when at least 80% of the control seeds had germinated (80-85 h) for *Lycopersicum esculentum*, 52 hours for *Cucumis sativus* and 48-50 hours for *Zea mays*, the experiment were carried out in triplicates [31].

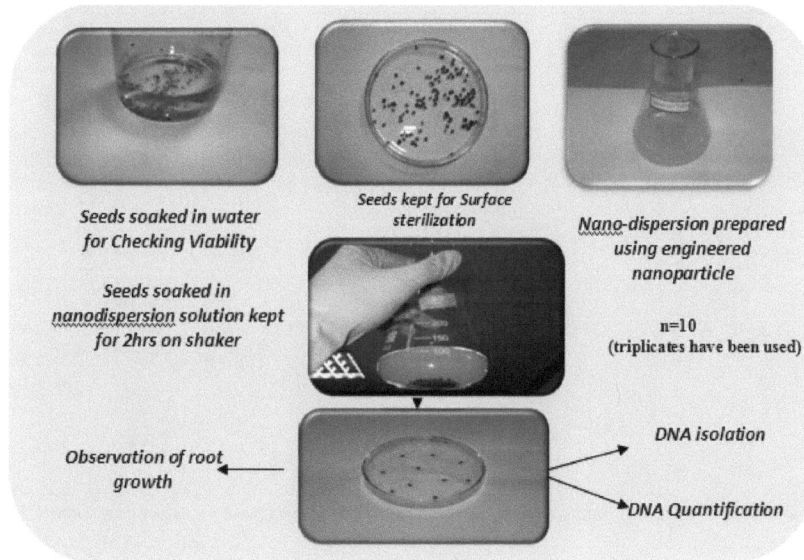

Figure 9: Exposure of nanoparticles dispersion with seed system.

Effect of Silver Nanoparticles on Root Length and Germination

There was a concentration dependent decrease in germination and root length for silver nanoparticles treated seeds of *L. esculentum*, *C. sativus* and *Z. mays*. These results are shown in Figs. **10**, **11** and **12** and Tables **2**, **3**, and **4**, respectively. The IC_{50} for the seed samples treated with silver nanoparticle was found to be 159.15 μg ml^{-1} for *L. esculentum*; 190.63 μg ml^{-1} for *C. sativus* and 237.9 μg ml^{-1} for *Z. mays*.

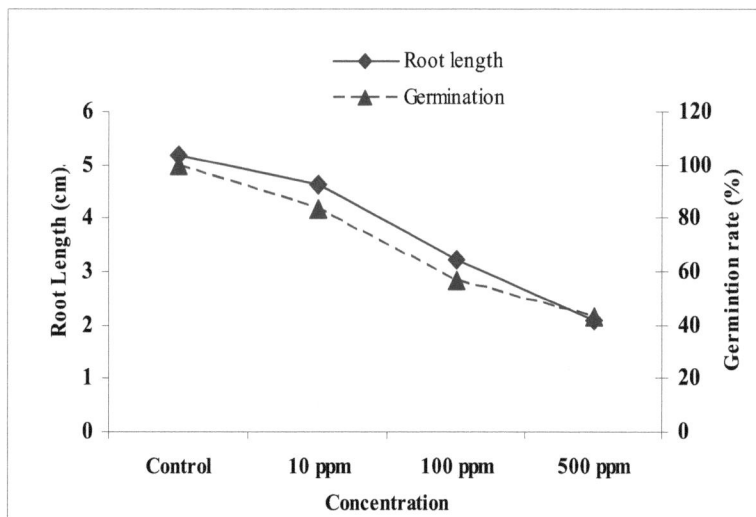

Figure 10: Graph showing root length and germination rate of *L. esculentum* exposed to silver nanoparticles.

Table 2: Germination results of *L. esculentum*.

Concentration (μg ml^{-1})				
(%)	Control	10	100	500
RSG	-	85	60	45
RRG	-	89.6	41.3	33.6
GI	-	74.6	23.5	14.5

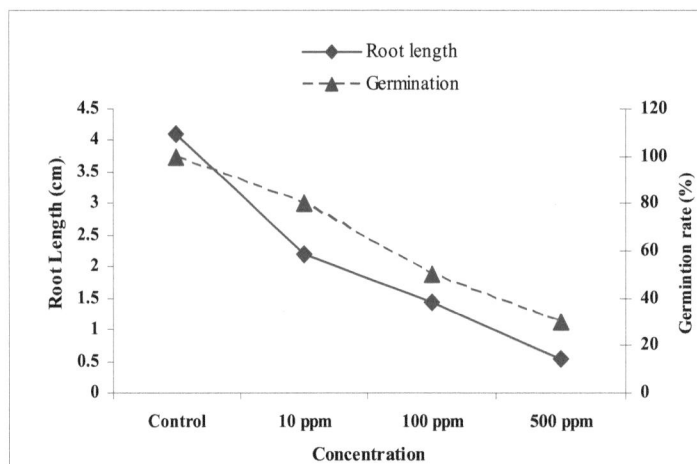

Figure 11: Graph showing root length and germination rate of *C. sativus* exposed to silver nanoparticles.

Table 3: Germination results of *C. sativus*.

Concentration (µg ml^{-1})				
(%)	Control	10	100	500
RSG	-	80	60	30
RRG	-	69.67	30.3	13.33
GI	-	55.7	18.17	67.73

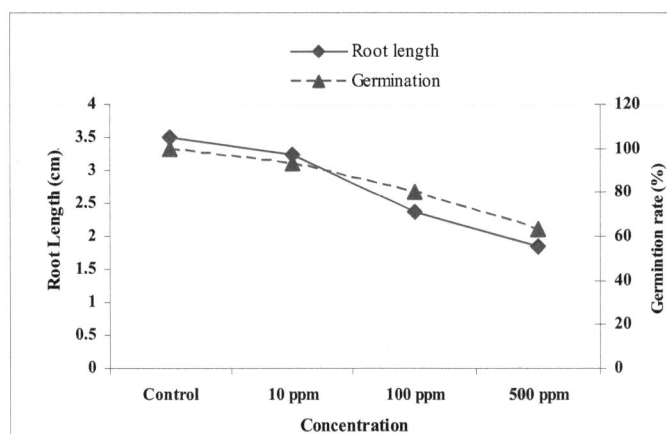

Figure 12: Graph showing root length and germination rate of *Z. mays* exposed to silver nanoparticles.

Table 4: Germination results of *Z. mays*.

Concentration (µg ml^{-1})				
(%)	Control	10	100	500
RSG	-	93.3	63.3	53.3
RRG	-	91.6	67	51
GI	-	85.4	42.46	27.3

Effect of Nano Zinc Oxide Particles on Root Length and Germination

There was a concentration dependent decrease in germination and root length for nano zinc oxide treated seeds of *L. esculentum*, *C. sativus* and *Z. mays*. These results are shown in Figs. **13**, **14** and **15** and Tables **5**,

6 and **7** respectively. The IC$_{50}$ for the seed samples treated with nano zinc oxide particles were found to be 261.4 µg ml^{-1} for *L. esculentum*; 238.95 µg ml^{-1} for *C. sativus*, and 363.54 µg ml^{-1} for *Z. mays*.

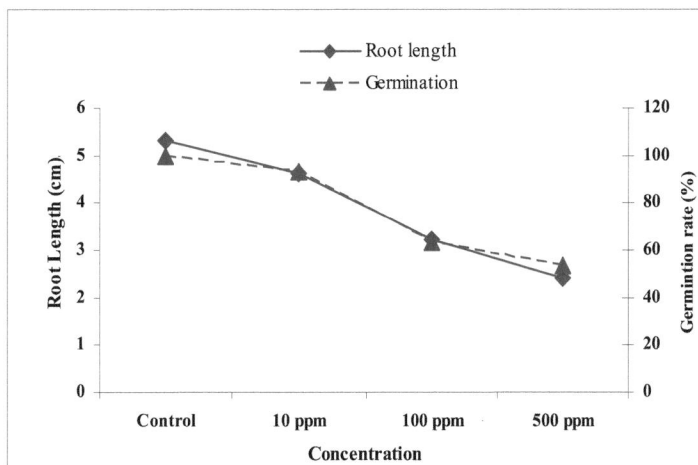

Figure 13: Graph showing root length and germination rate of *L. esculentum* exposed to nano zinc oxide.

Table 5: Germination results of *L. esculentum*.

Concentration (µg ml^{-1})				
(%)	Control	10	100	500
RSG	-	83.33	56.67	43.33
RRG	-	89.61	46.63	44.66
GI	-	96.16	123.2	80.7

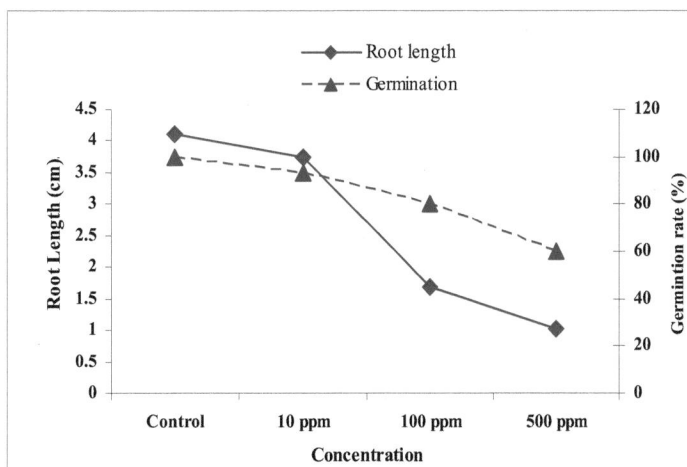

Figure 14: Graph showing root length and germination rate of *C. sativus* exposed to nano zinc oxide.

Table 6: Germination results of *C. Sativus*.

Concentration (µg ml^{-1})				
(%)	Control	10	100	500
RSG	-	93.33	80	60
RRG	-	87.8	41.1	26.57
GI	-	106.7	197.6	227.1

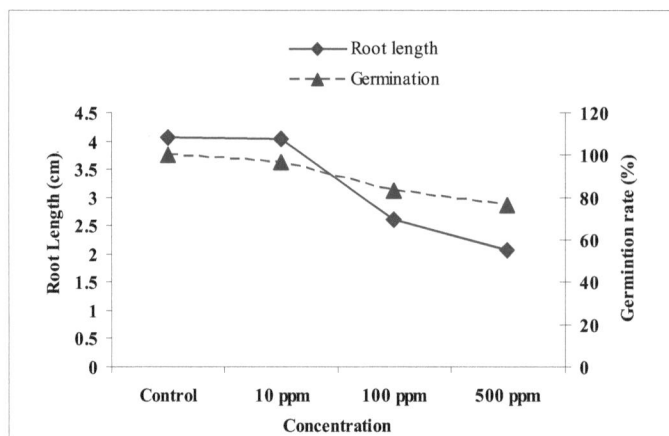

Figure 15: Graph showing root length and germination rate of *Z. mays* exposed to nano zinc oxide.

Table 7: Germination results of *Z. Mays.*

Concentration (μg ml⁻¹)				
(%)	Control	10	100	500
RSG	-	93.3	63.3	53.3
RRG	-	91.63	58.07	50.07
GI	-	101.9	91.23	106.1

IN VIVO STUDIES: SEED EXPERIMENT

Seedling Exposure Using Phytagel

In the present study plants were used to access the soil ecotoxicity and bioavailability of silver nanoparticles. *Cucumis sativus* and *Zea mays* were selected as the test species because of their commercial importance, less germination time, sensitive to toxicity, routine use in phytotoxicity tests and their importance as food plants. *Lycopersicum esculentum* was not selected as test species for *in vivo* studies because of very small size and high germination time. Most nanoparticles are hardly water soluble so a plant agar test has been employed to avoid precipitation of nanoparticles and to distribute nanoparticles evenly in test species.

Preparation of Culture Dispersion

Culture dispersion of nanoparticles was achieved by adding phytagel powder (Sigma-Aldrich, USA, melting point 90°C) and the needed amount of nanoparticles to deionized water. The dispersions were sufficiently shaken after sonication to break up agglomerates. Each nanoparticle treatment concentration was prepared separately without dilution to avoid agglomeration in the dispersion and sonicated for 30 min. Agar was melted separately in deionized water. Nano dispersion was added to the melted agar at an optimum temperature of 40-45°C to avoid temperature effect on nanoparticles. The concentration of nanoparticles ranged from 0 to 500 μg ml⁻¹. The agar culture media have the advantage of easy dispersion of nanoparticles without precipitation.

Acute Toxicity Test using Plant Agar Method

Seeds were sterilized in a 10% sodium hypochlorite solution for 10 min, rinsed thoroughly with deionized water and subsequently placed in wet cotton at a controlled temperature of 25±1°C in the dark. After 24 h, seeds were checked for the germination and seeds that had sprouted were used in the test. The toxicity tests were conducted in beakers. Each test unit contained 30 ml of 1.5 % of agar media with a specific concentration of nanoparticles and immediately hardened in a freezer to avoid the possible precipitation of nanoparticles. Test plant seedlings were placed just above the surface of the agar in the test units (Fig. **16**).

The test units were kept for incubation at a controlled temperature of 25±1° C in dark. The exposure period was 72 h. After an incubation period of 3 days, the plants were separated from the agar media, and seedling growth was measured. IC$_{50}$ was calculated using probit analysis [32].

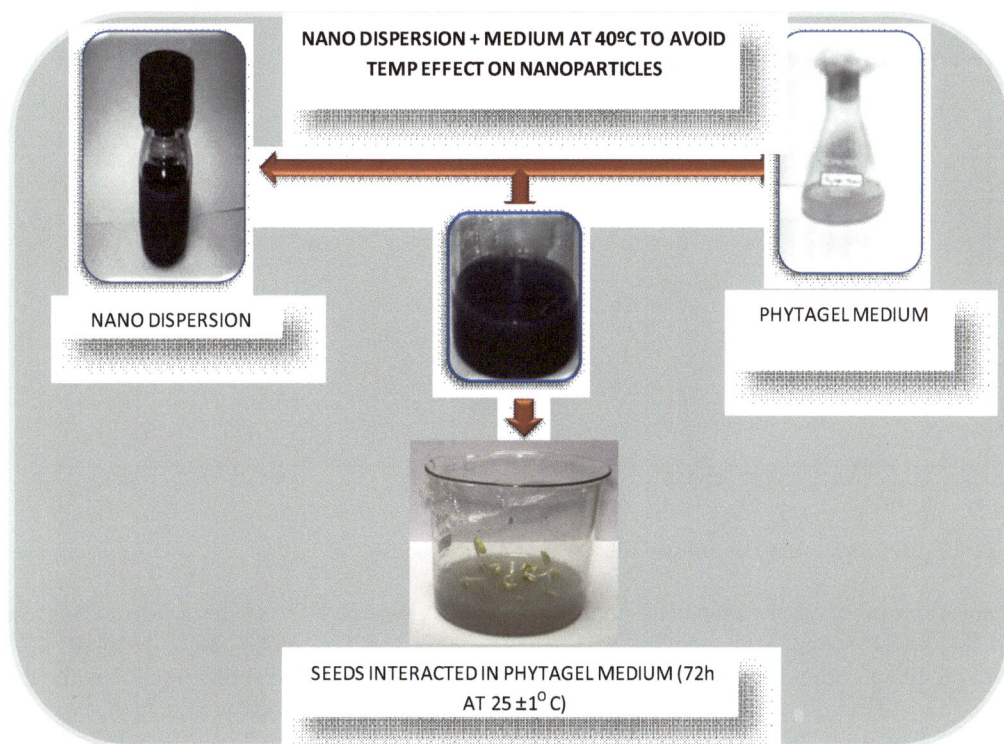

Figure 16: Interaction of nanoparticles in phytagel with seed system.

Effect of Silver Nanoparticles on Root and Shoot of C. sativus

The effect of silver nanoparticles showed decrease on germination, root and shoot length with increasing the concentration of the silver nanoparticles. There was no growth of root and shoot at 500 μgml^{-1} silver nanoparticles treated samples for *C. sativus*. The root length for the sample treated with 10, 100 and 500 μgml^{-1} were found to be 2.67, 0.53 and 0 cm respectively, shoot length showed 2.1, 0.59 and 0 cm respectively; control seeds showed 4.5 cm for root and 4.4 for shoot for *C. sativus*; germination rate was found to be 100 % for control. The root germination for the sample treated with 10, 100 and 500 μgml^{-1} *C. sativus* showed 23.3, 57, and 10 % and shoot showed 90, 30 % and no germination respectively. There was minimum growth of root and shoot of 500 μg ml^{-1} silver nanoparticles treated samples of *Z. mays*. As compare to *C. sativus*, *Z. mays* showed growth at higher concentration; indicating that nanoparticles toxicity was more prominent on *C. sativus*.

The root length for the sample treated with 10, 100 and 500 μgml^{-1} silver nanoparticles were 4.2, 3.7 and 3.1 cm and shoot length showed 4.7, 3.1, 2.1 cm respectively; control seeds showed 5.4 cm root and 5.7 cm for shoot. For *Z. mays* root germination showed 83.6, 60 and 30% and shoot germination showed, 80, 40, 20% germination respectively. In all the samples control seeds showed 100% germination.

FT-IR ANALYSIS FOR CONTROL AND NANOPARTICLE TREATED SAMPLES

FT-IR spectroscopy was done to analyze various chemical and conformational changes occurring in the cellular components. The FTIR spectrum obtained for the uninteracted root cells of *A. cepa* revealed the following major characteristic peaks: 3420.15 cm^{-1} band predominantly due to amide-I stretching, 1633.16 cm^{-1} due to amide-II stretching and 1026.6 cm^{-1} due to carbohydrates (Fig. **17**).

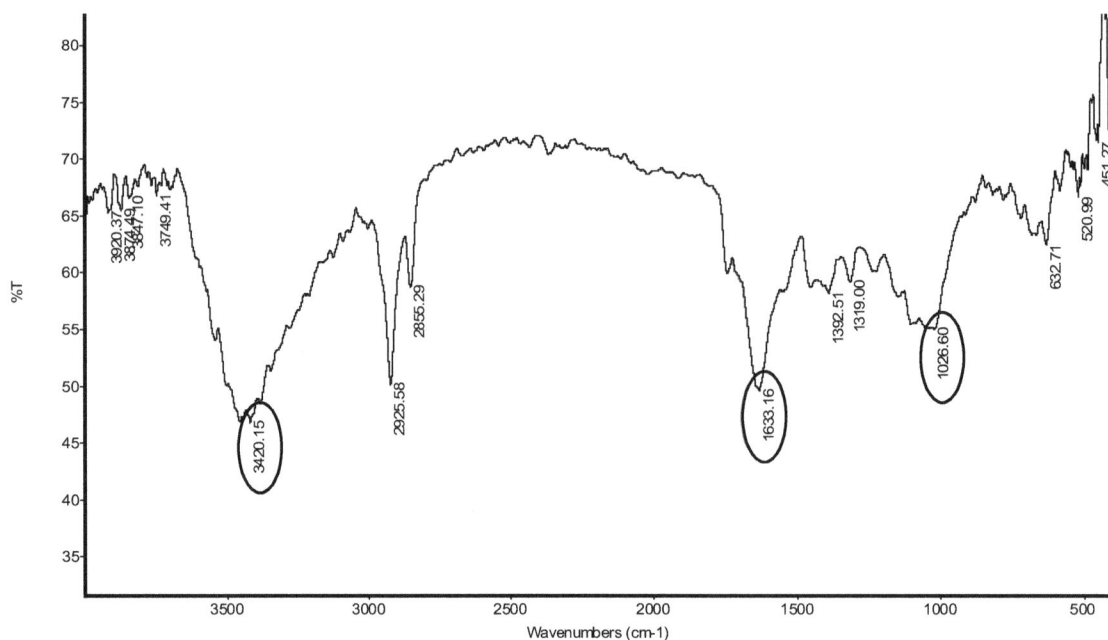

Figure 17: FT-IR spectrum of *A. cepa* treated with distilled water.

FT-IR spectrum of silver nanoparticles interacted samples revealed shift in the following regions: for amide-I 3420.5-3445.5 cm^{-1}, for amide-II 1633.1634.42 cm^{-1} and for carbohydrates 1026-1034.63 cm^{-1} (Fig. **18**). FT-IR spectrum of the nano zinc oxide interacted samples revealed shift in the following regions: for amide-I in 3420.15-3409.25 cm^{-1}, for amide-II in 633.16-1639.17 cm^{-1}, and for carbohydrates in 1026.6-1105.1 cm^{-1} (Fig. **19**). This conformational change depicts the interaction of nanoparticles with the test samples. The presence of 447.49 cm^{1} peak lies in the fingerprint region which corresponds to nano zinc oxide particles.

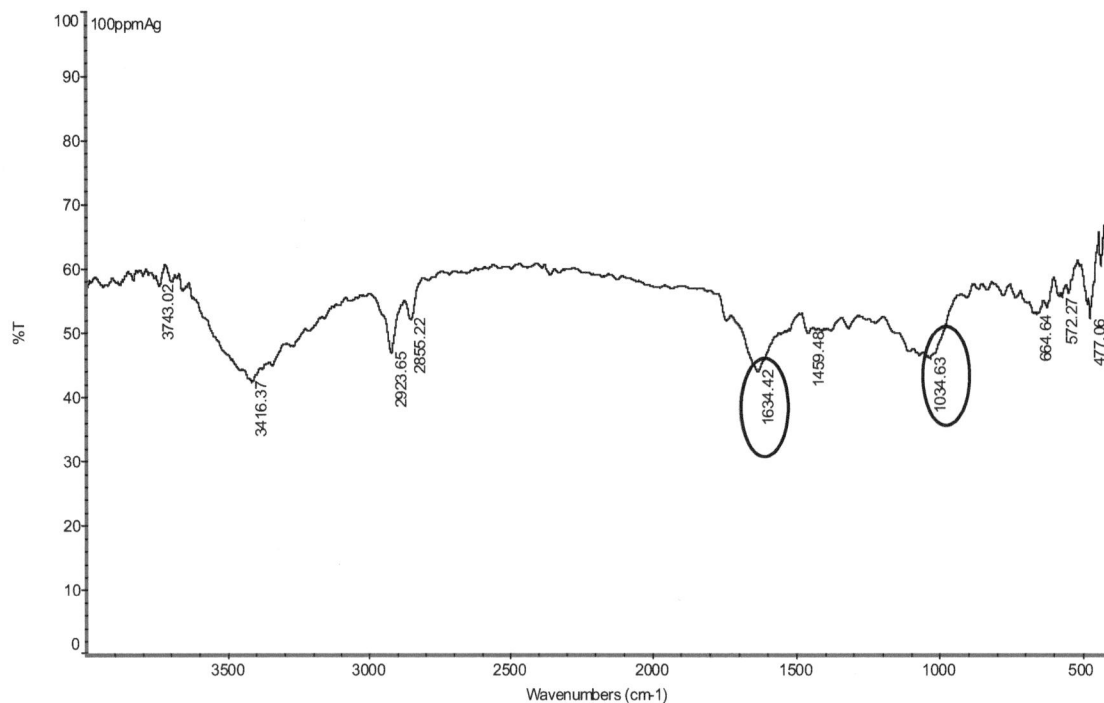

Figure 18: FT-IR spectrum of *A. cepa* treated with silver nanoparticles.

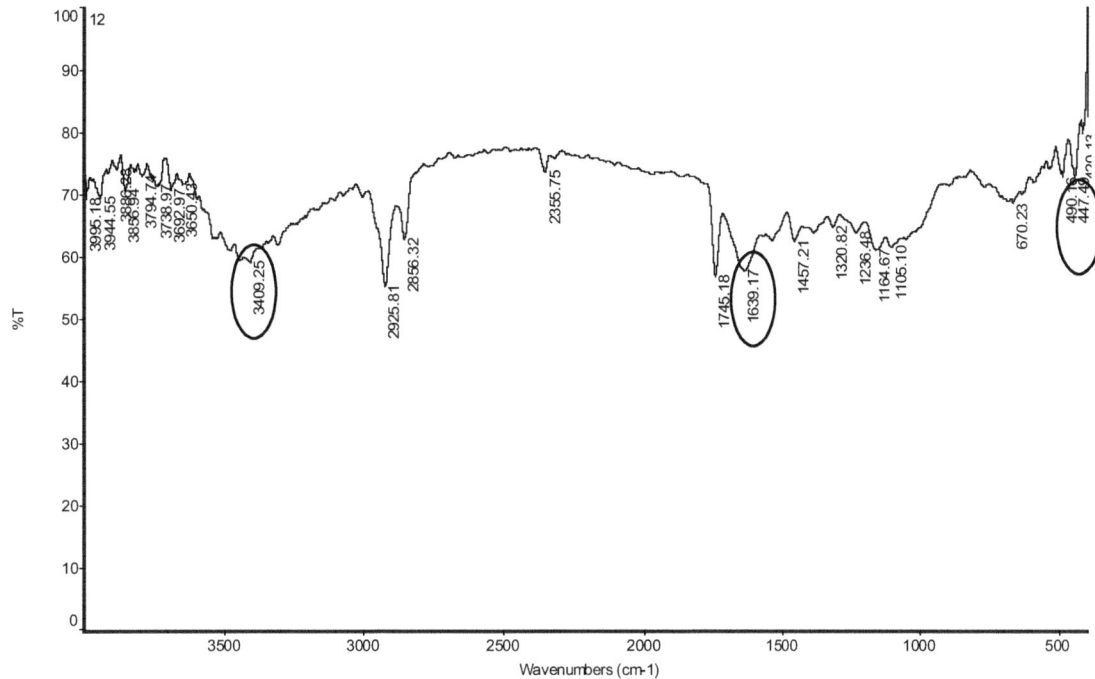

Figure 19: FT-IR spectrum of *A. cepa* treated with nano zinc oxide particles.

FT-IR Analysis for L. esculentum

The characteristic peaks for the FT-IR spectrum obtained for the uninteracted samples revealed the following major characteristics peaks: amide-I (3394.65 cm^{-1}), amide-II (1642.16 cm^{-1}), carbohydrate (1024.39 cm^{-1}), and aldehyde (2925.81 cm^{-1}) (Fig. **20**). Deformation were seen in amide-I (3394.65-3427.98), amide-II (1642.16 -1634.42 cm^{-1}) carbohydrate regions in silver nanoparticles treated samples (Fig. **21**).

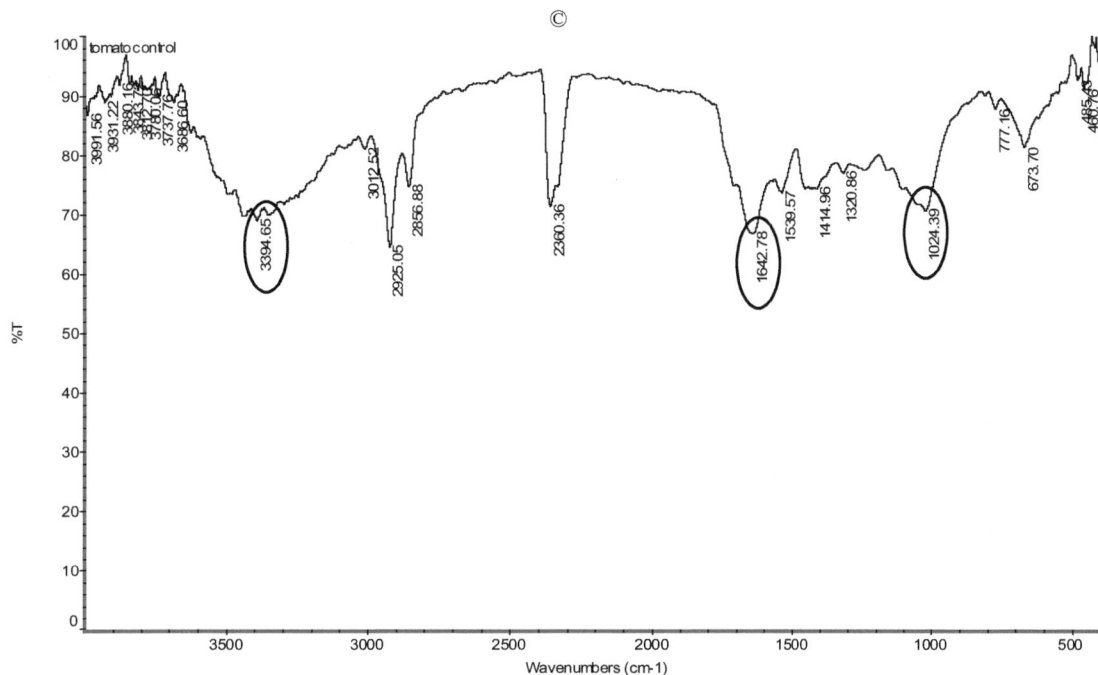

Figure 20: FT-IR spectrum of *L. esculentum* treated with distilled water.

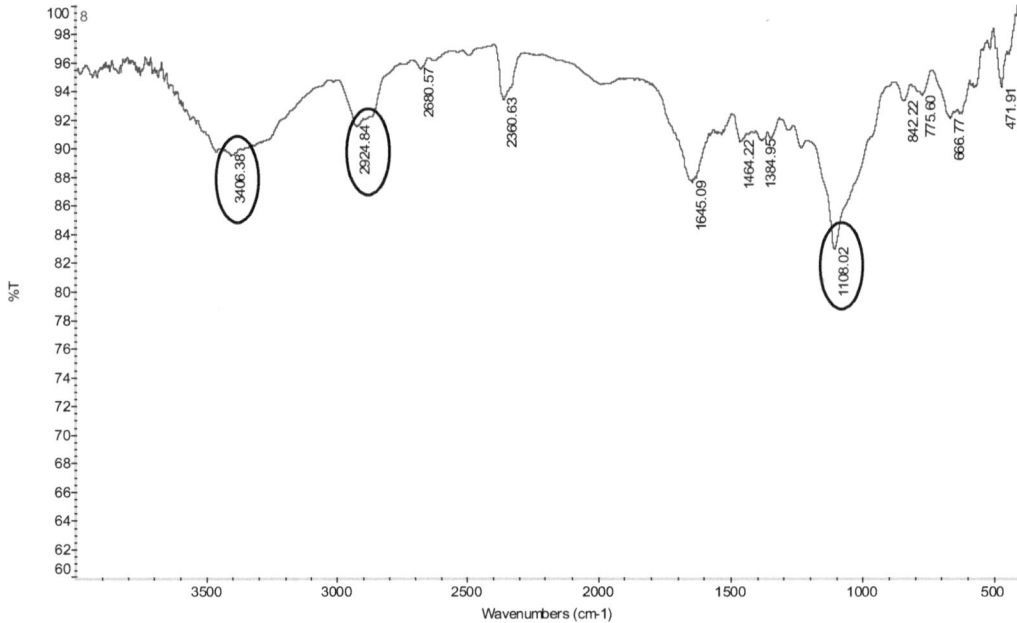

Figure 21: FT-IR spectrum of *L. esculentum* treated with nano zinc oxide particles.

Nano zinc oxide particles treated samples showed deformation in the entire region as follows: band shift was found at, amide-I (3409-3122.35 cm^{-1}), amide-II (1642.16-1645.09 cm^{-1}), carbohydrate (1024.39-1108.02 cm^{-1}) and aldehyde (2925.81-2924.4 cm^{-1}) (Fig. **22**).

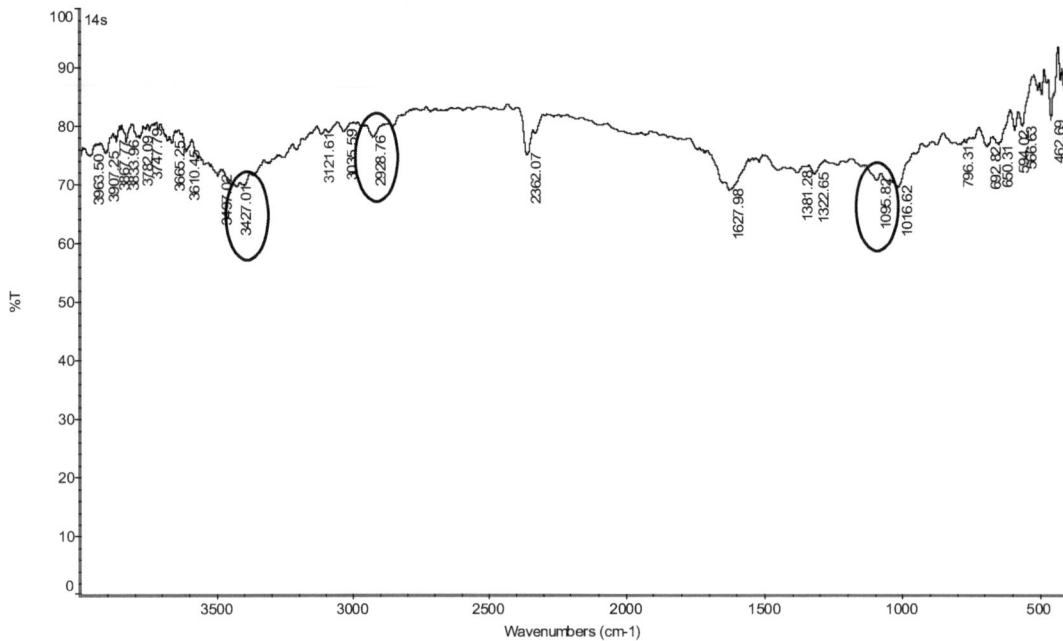

Figure 22: FT-IR spectrum of *L. esculentum* treated with silver nanoparticles.

FT-IR Analysis for C. sativus

The FT-IR spectrum obtained for the uninteracted samples revealed the following major characteristics peaks: amide-I (3409.25 cm^{-1}), amide-II (1639.17 cm^{-1}), carbohydrate (1025.52 cm^{-1}), and aldehyde (2925.81 cm^{-1}) (Fig. **23**). Deformation was seen in amide-I (absence of peak), amide-II (1639.17 -1637.38

cm^{-1}) and carbohydrate regions in SNP treated samples (Fig. **24**). Nano zinc oxide particles treated samples showed deformation in the entire region as follows: band shift was found at, amide-I (3409-3122.35 cm^{-1}), amide-II (1639.17-1627.98 cm^{-1}), carbohydrate (1025.52-1095.82 cm^{-1}) and aldehyde (2925.81-2928.76 cm^{-1}) (Fig. **25**).

FT-IR Analysis for Z. mays

The FTIR spectrum obtained from the un-interacted samples revealed the following major characteristics peaks: amide-I (3356.23 cm^{-1}), amide-II (1633.88 cm^{-1}), carbohydrate (1020.43 cm^{-1}), and for aldehyde group (2927.82 cm^{-1}) (Fig. **26**). Deformation were seen in amide-I (3356.23-3440.72 cm^{-1}), amide-II (1633.88-1627.98 cm^{-1}), carbohydrate regions (absence of peak) in silver treated samples (Fig. **27**).

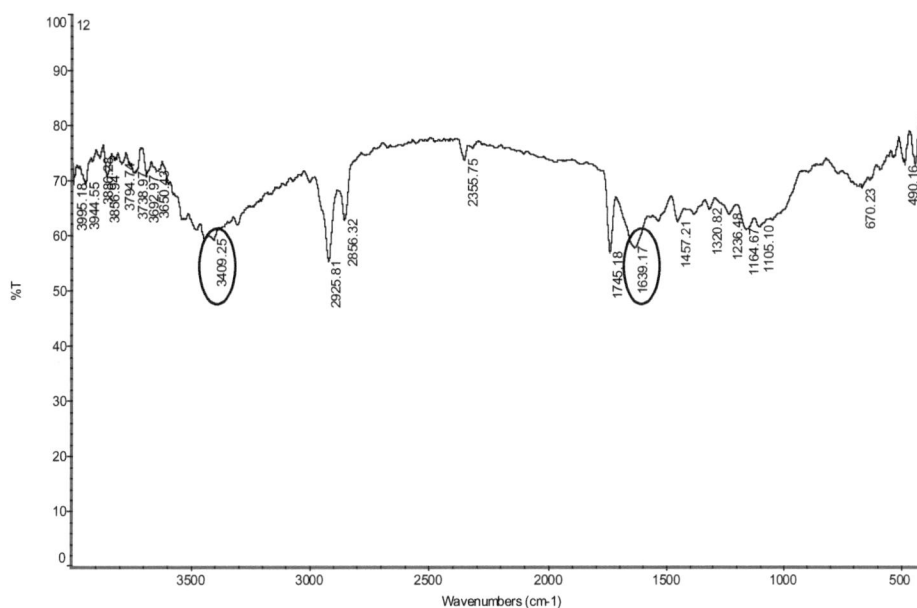

Figure 23: FT-IR spectrum of *C. sativus* treated with distilled water.

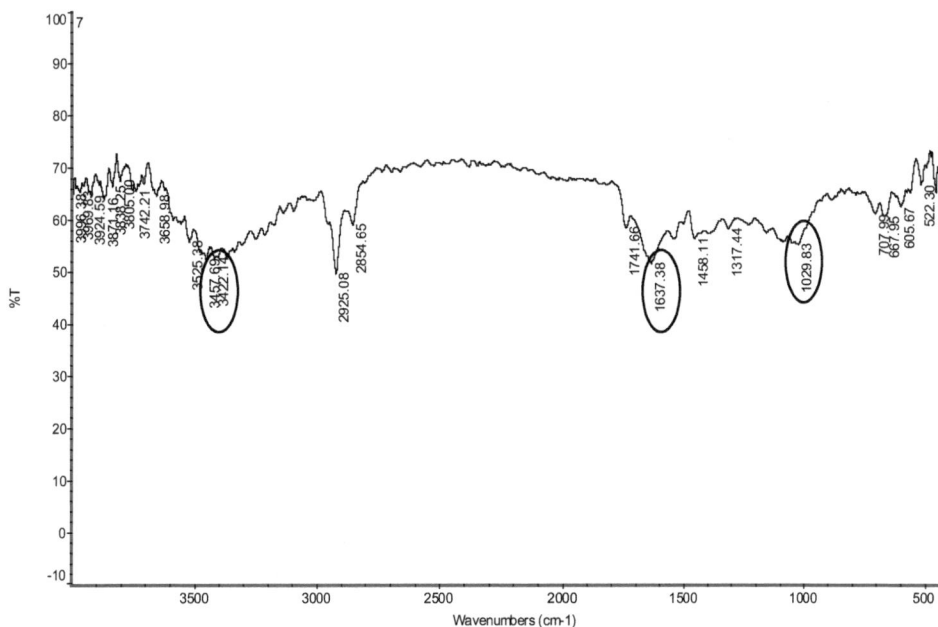

Figure 24: FT-IR spectrum of *C. sativus* treated with silver nanoparticles.

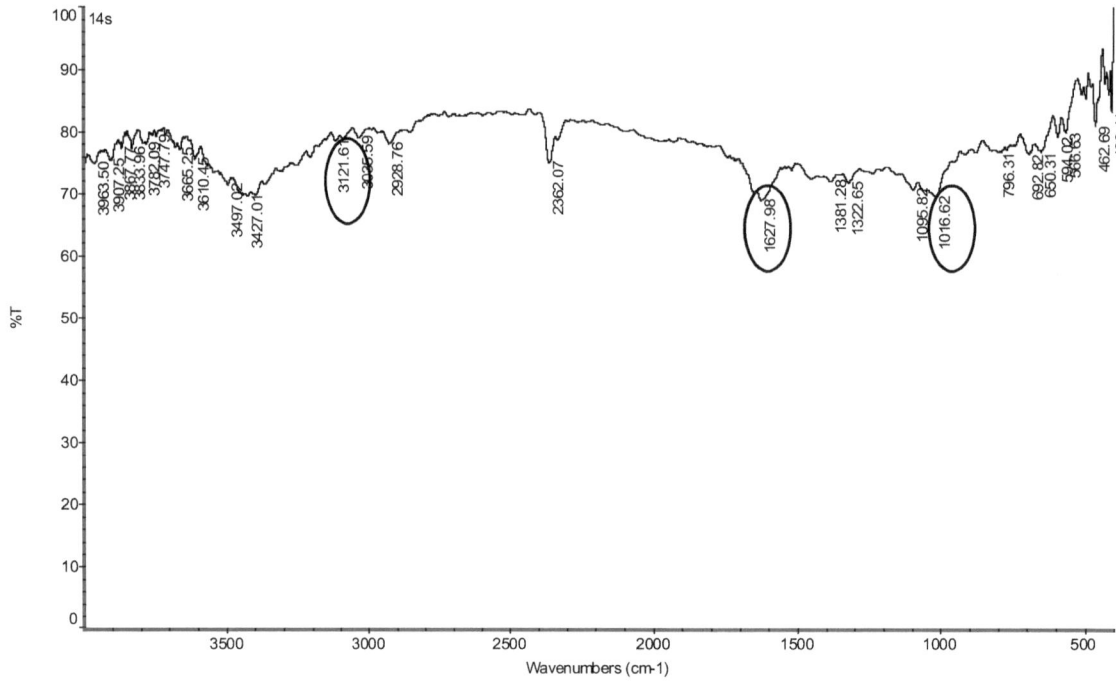

Figure 25: FT-IR spectrum of *C. sativus* treated with nano zinc oxide particles.

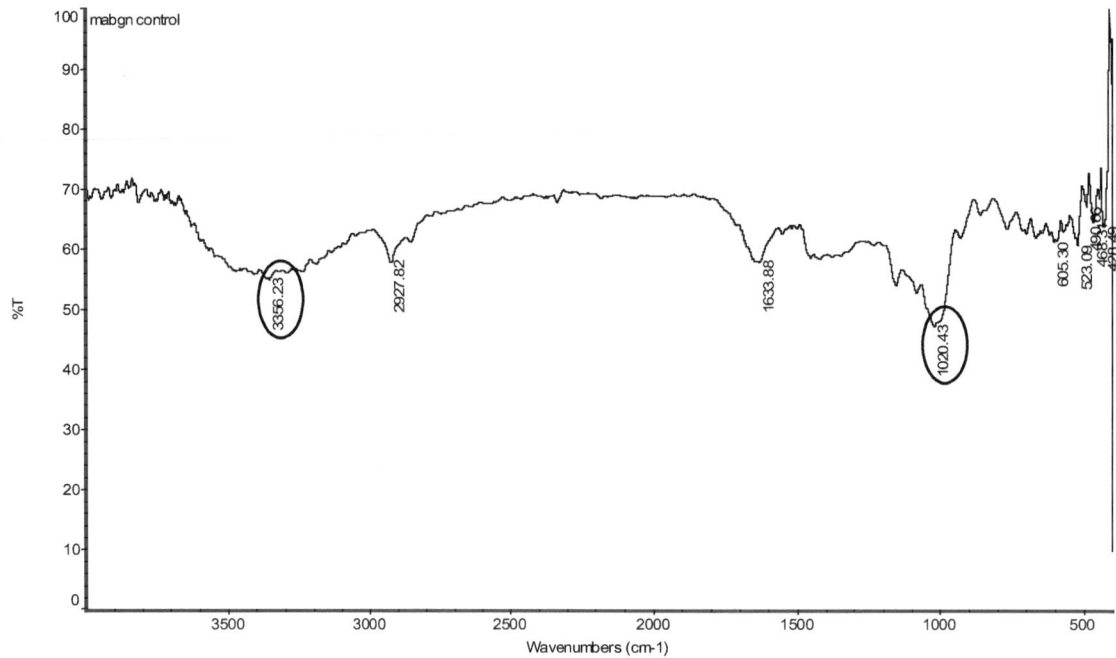

Figure 26: FT-IR spectrum of *Z. mays* treated with distilled water.

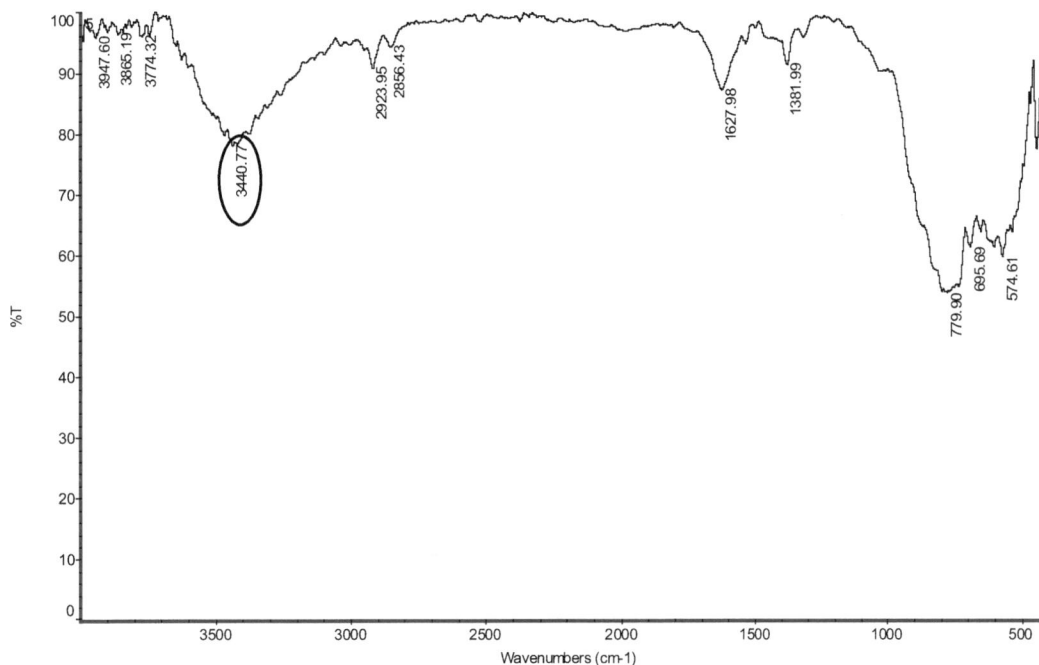

Figure 27: FT-IR spectrum of *Z. mays* treated with silver nanoparticles.

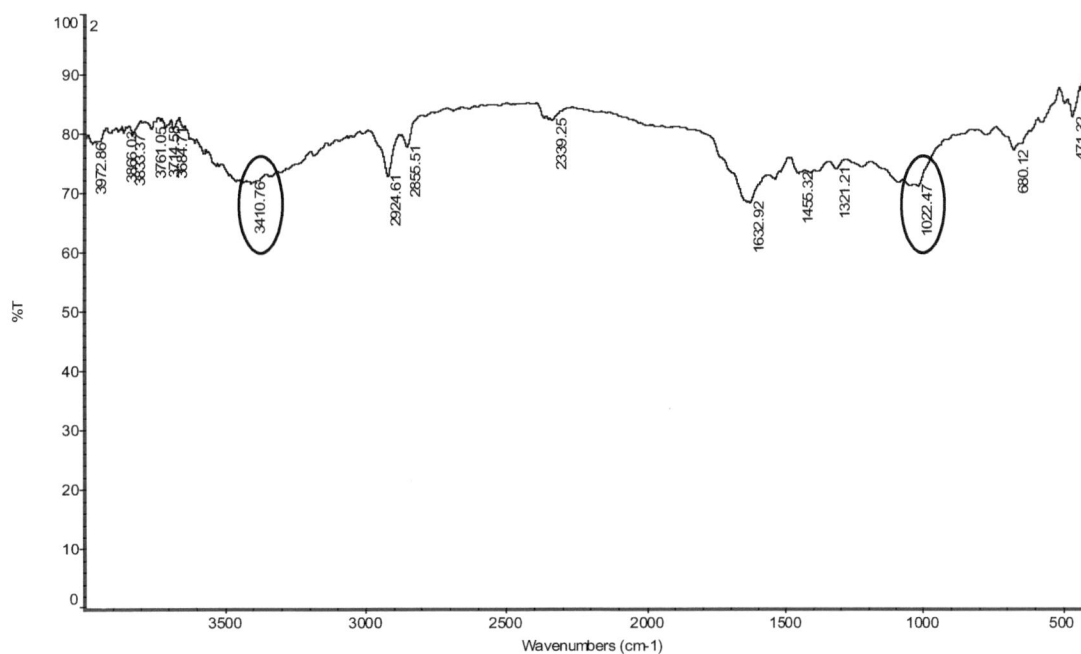

Figure 28: FT-IR spectrum of *Z. mays* treated with nano zinc oxide particles.

Nano zinc oxide particles treated samples showed deformation in the entire region as follows: band shift was found to be, amide-I (3356-3410.76 cm^{-1}), amide-II (1633.88-1632.92 cm^{-1}), carbohydrate (1025.52-1095.82 cm^{-1}) and aldehyde (2927.82-2924.61 cm^{-1}) (Fig. **28**).

ADSORPTION STUDIES

Adsorption studies were done by washing the exposed seeds thrice in distilled water followed by quantification of nanoparticles in the supernatant using atomic absorption spectroscopy (Varian-AA 240).

This study showed the metal adsorbed onto the surface of the seeds. This could be due to large surface site densities (positively or negatively charged sites) and cation exchange capacities (negatively charged sites only) on to the surface of nanoparticles and seeds and also because of the extent of metal adsorption depends on the total metal concentration and the pH. (Tables **8** and **9**)

Table 8: Adsorption studies for silver nanoparticles treated seed samples.

Seed Systems	Concentration ($\mu g\ ml^{-1}$)		
	10	**100**	**500**
C. sativus	2.87 ± 1.98	8.19 ± 2.78	19.06 ± 1.31
Z. mays	3.12 ± 1.18	9.9 ± 0.92	17.6 ± 2.35

Table 9: Adsorption studies for nano zinc oxide treated seed samples.

Seed Systems	Concentration ($\mu g\ ml^{-1}$)		
	10	**100**	**500**
C. sativus	2.07 ± 1.91	9.39 ± 1.78	17.09 ± 1.38
Z. mays	2.9 ± 2.03	6.7 ± 0.98	15.6 ± 2.31

ACCUMULATION OF NANOPARTICLES IN ROOTS

To determine silver and zinc oxide nanoparticles accumulation in plant root tissue after 72-82 hours, all plants were washed thoroughly with distilled water to remove the test medium. Total silver concentrations are reported here as a weight percentage on a dry plant tissue basis. 0.1g of sample was dissolved in 2 ml of conc. HNO_3 and 2 ml of deionized water at 90°C for 2h and the solution was filtered through a glass frit; it was made up to a specified volume in a volumetric flask. The concentration was measured using an atomic absorption spectrophotometer.

ACCUMULATION STUDIES FOR SILVER NANOPARTICLES TREATED SEED SAMPLES

Accumulation studies showed maximum uptake by *C. sativus* than *L. esculentum* and *Zea mays*. It is due to the increased surface area of root in *C. sativus*. SNP uptake in *C. sativus* was 1.67, 11.4 and 15.9 $\mu g\ ml^{-1}$ for 10, 100 and 500 $\mu g\ ml^{-1}$ treated samples respectively; *L. esculentum* showed 1.44, 8.5, 10.2 $\mu g\ ml^{-1}$ for 10, 100 and 500 $\mu g\ ml^{-1}$; and for *Z. mays* was 1.2, 7.23, and 9.9 $\mu g\ ml^{-1}$ at 10, 100 and 500 $\mu g\ ml^{-1}$ treated samples respectively (Table **10**).

Table 10: Accumulation of SNP treated seed samples.

Seed Systems	Concentration ($\mu g\ ml^{-1}$)		
	10	**100**	**500**
C. sativus	1.67 ± 0.98	11.4 ± 1.01	15.9 ± 1.03
L.esculentum	1.44 ± 1.02	8.5 ± 0.89	10.2 ± 1.02
Z. mays	1.2 ± 1.08	7.23 ± 0.92	9.9 ± 0.35

ACCUMULATION STUDIES FOR NANO ZINC OXIDE TREATED SEEDS SAMPLES

Accumulation of zinc oxide nanoparticles in roots of seeds system showed maximum zinc accumulation in *C. sativus* and least in *Z. mays*. Zinc accumulated in *C .sativus* for 10, 100 and 500 $\mu g\ ml^{-1}$ treated samples showed 2.04 12.4, and 16.7 $\mu g\ ml^{-1}$ respectively; for *L. esculentum* it was 1.53, 9.01 and 12.09 $\mu g\ ml^{-1}$ respectively; and for *Z. mays* it was 1.4, 7.23 and 11.29 $\mu g\ ml^{-1}$ respectively (Table **11**).

Table 11: Accumulation of nano zinc oxide treated seed samples.

Seed Systems	Concentration (µg ml^{-1})		
	10	100	500
C. sativus	2.04 ± 1.98	12.4 ± 1.01	16.7 ± 1.73
L .esculentum	1.53 ± 2.12	9.01 ± 0.98	12.09 ± 1.42
Z. mays	1.4 ± 1.08	7.23 ± 0.92	11.29 ± 0.78

CONCLUSION

Silver nanoparticles were found to be more toxic than nano zinc oxide particles to *A. cepa* and seed systems. The findings also suggest that plants being an important component of the environmental and ecological systems should be included when evaluating the overall fate, transport, and exposure pathways of nanoparticles in the environment. Thus, before dumping a huge amount of nanomaterials into the environment, there is a need to investigate the solubility and degradability of engineered nanoparticles in soil and water and to establish baseline information on their biosafety.

LIST OF ABBREVIATIONS

AAS, Atomic Absorption Spectrophotometer; AFM, Atomic Force Microscope; FT-IR, Fourier Transform Infrared Spectroscopy; KHz, Kilo hertz; M, Molar; MI, Mitotic index; MN, Micronucleus; ROS, Reactive Oxygen Species; SEM, Scanning Electron Microscope; TC, Total Cells; TDC, Total Dividing Cells; TEM, Transmission Electron Microscope; XRD, X-Ray Diffraction

REFERENCES

[1]   Osterberg R, Persson D, Bjursell G. The condensation of DNA by chromium (III) ions. J Biomol Struct Dyn 1984; 2: 285-290.
[2]   The Royal Society and Royal Academy of Engineering, UK. Nanoscience and Nanotechnology, Opportunities and Uncertainties. UK: The Royal Society: 2004.
[3]   Munzuroglu O, Geckil H. Effects of metals on seed germination, root elongation, and coleoptile and hypocotyl growth in *Triticum aestivum* and *Cucumis sativus*. Arch. Environ Contam Toxicol 2002; 43: 203-213.
[4]   Oberdorster G, Oberdorster E, Oberdorster J. Nanotoxicology: an emerging discipline evolving from studies of ultrafine particles. Environ Health Perspect 2005; 113: 823-839.
[5]   Anastasio C, Martin ST. Atmospheric nanoparticles. Rev Miner Geochem 2001; 44 293-349.
[6]   Nel A, Xia T, Madler L, Li N. Toxic potential of materials at the Nanolevel. Science 2006; 311: 622-627.
[7]   Toxicological profile for Silver. Contract No: 205-88-0608. Prepared by Clement international corporation, U.S. Public Health Service, ATSDR/TP-90-24. ATSDR (Agency for toxic substances and Disease Registry) 1990.
[8]   Maynard AD, Aitken RJ, Butz T, *et al.* Safe handling of nanotechnology. Nature 2006; 444: 267-269.
[9]   Chen X, Schluesener HJ. Nanosilver: a nanoproduct in medical application. Toxicol Lett 2008; 176 (Suppl 1): 1-12.
[10]   Tripathy A, Chandrasekran N, Raichur AM and Mukherjee A. Antibacterial applications of silver nanoparticles synthesized by aqueous extract of Azadirachta indica (Neem) leaves. J Biomed Nanotech 2008; 4: 1-6.
[11]   USEPA. European Agency for Safety and Health Report, 2009.
[12]   U.S. Environmental Protection Agency Nanotechnology White Paper e External Review Draft. Available from: http://www.epa.gov/osa/pdfs/ EPA_nanotechnology_white_paper_external_review_draft_12-02-2005, [cited Dec 2, 2005].
[13]   Wiesner MR, Lowry GV, Alvarez P, Dionisiou D, Biswas P. Environ Sci Technol 2006; 15: 4336-4345.
[14]   Xia T, Kovochich M, Brant J, *et al.* Comparison of the abilities of ambient and manufactured nanoparticles to induce cellular toxicity according to an oxidative stress paradigm. Nano Lett 2006; 6: 1804-1807.
[15]   Hsin Y, Chen C, Huang S, Shih T, Lai P, Chueh PJ. The apoptotic effect of nanosilver is mediated by a ROS- and JNK-dependent mechanism involving the mitochondrial pathway in NIH3T3 cells. Toxicol Lett 2008; 179: 130-139.
[16]   Franklin NM, Rogers NJ, Apte SC, Batley GE, Gadd GE, Casey PS. Comparative toxicity of nanoparticulate ZnO, bulk ZnO, and ZnCl to a freshwater microalga (*Pseudokirchneriella subcapitata*): the importance of particle solubility. Environ Sci Technol 2007; 41: 8484-8490.

[17]   USEPA, Nanotechnology White Paper, *Science Policy Council*, Washington, DC, Available from: www.epa.gov/osainter/../nanotech/epa-nanotechnology-whitepaper-0207.pdf [cited Feb 15 2007].

[18]   Zhu H, Han J, Xiao JQ, Jin Y. Uptake, translocation and accumulation of manufactured iron oxide nanoparticles by pumpkin plants. J Environ Monit 2008; 10: 713-717.

[19]   Leme DM, Marin-Morales MA. Chromosome aberration and micronucleus frequencies in *Allium cepa* cells exposed to petroleum polluted water - a case study. Mutat Res 2008; 650: 80-86.

[20]   Grant WF. Chromosome aberration assay in *Allium cepa*. Mutat Res 1982; 99: 273-291.

[21]   Fiskejo G. The Allium test as a standard in environmental monitoring. Hereditas 1985; 102: 99-112.

[22]   Lu CM, Zhang CY, Wu JQ, Wen GR, Tao MX. Research of the effect of nanometer materials on germination and growth enhancement of glycine max and its mechanism. Soybean Science 2002; 21: 168.

[23]   Hong FS, Zhou J, Liu C, Yang F, Wu C, Zheng L, Yang P. Effect of nano-TiO$_2$ on photochemical reaction of chloroplasts of spinach. Biol Trace Elem Res 2005; 105: 269.

[24]   Yang F, Hong FS, You WJ, Liu C, Gao FQ, Wu C, Yang P. Influences of nano-anatase TiO$_2$ on the nitrogen metabolism of growing spinach. Biol Trace Elem Res 2006; 110: 179-190.

[25]   Zheng L, Hong FS, Lu SP , Liu C. Effect of nano-TiO2 on strength of naturally aged seeds and growth of spinach. Biol Trace Elem Res 2005; 104: 83.

[26]   Kumari M, Mukherjee A, Chandrasekaran N. Genotoxicity of Silver Nanoparticles. Sci Total Env 2009; 407: 5243-5246.

[27]   Lin D, Xing B. Root uptake and Phytotoxicity of ZnO nanoparticles, Environ Sci Technol 2008; 42: 5580-5585.

[28]   Wang HH, Wick RL, Xing BS. Toxicity of nanoparticles and bulk ZnO, Al$_2$O$_3$ and TiO$_2$ to the nematode *Caenorhabditis elegans*. Environ Pollut 2009; 157 (Suppl 2): 1171-1177.

[29]   Fiskesjo G. Allium test for screening chemicals; evaluation of cytologic parameters. In: Wang W, Gorsuch JW, Hughes JS, editors. Plants for environmental studies. Boca Raton, New York: CRC Lewis Publishers 1997; pp. 308.

[30]   Ecological Effects Test Guidelines (OPPTS 850.4200), Seed Germination/Root Elongation Toxicity Test. Available from:
http://www.epa.gov/opptsfrs/publications/OPPTS_Harmonized/850_Ecological_Effects_Test_Guidelines/Drafts/850
-4200.pdf. U.S. Environmental Protection Agency 1996.

[31]   Yang L, Watts DJ. Particle surface characteristics may play an important role in phytotoxicity of alumina nanoparticles. Toxicol Lett 2005; 158:122-132.

[32]   Data Quality Objectives Decision Error Feasibility Trials *(DQO/DEFT)*, Version 4.0, EPA QA/G-4D, Washington, D.C, USEPA, 1994.

[33]   Bakare AA, Mosuro AA, Osibanjo O. Effect of simulated leachate on chromosomes and mitosis in roots of *Allium cepa* (L). J Environ Biol 2000; 21: 263.

[34]   Tolbert PE, Shy CM, Allen JW. Micronuclei and other nuclear abnormalities in buccal smears, method and development. Mutat Res 1992; 271: 69-71.

[35]   Wierzbicka M. Lead in the apoplast of *Allium cepa* L. root tips - ultrastructural studies. Plant Sci 1998; 133: 105-119.

[36]   Corredor E, Testillano PS, Coroanado MJ, *et al.* Nanoparticle penetration and transport in living pumpkin plants: *in situ* subcellular identification. BMC Plant Biol 2009; 9: 45.

[37]   Lee WM, An Y, Yoon H, Kweon H. Toxicity and Bioavailability of copper nanoparticles to the terrestrial plants bean (*Triticum aestivum*): plant agar test for water-insoluble nanoparticles. Env Toxicol Chem 2008; 27: 1915-1921.

CHAPTER 3

Cell Life Cycle Effects of Bare and Coated Superparamagnetic Iron Oxide Nanoparticles

Morteza Mahmoudi[1,2*], Sophie Laurent[3] and W. Shane Journeay[4,5]

[1]National Cell Bank, Pasteur Institute of Iran, Tehran, Iran; [2]Institute for Nanoscience and Nanotechnology, Sharif University of Technology, Tehran, Iran; [3]Department of General, Organic, and Biomedical Chemistry, NMR and Molecular Imaging Laboratory, University of Mons, Belgium; [4]Nanotechnology Toxicology Consulting and Training, Inc, Nova Scotia, Canada and [5] Faculty of Medicine, Dalhousie Medical School, Dalhousie University, Halifax, Nova Scotia, Canada

Abstract: Due to the hopeful potential of nanoparticles in medicine, they have attracted much attention for various applications such as targeted drug/gene delivery, separation or imaging. Interaction of NPs with the biological environment can lead to a wide range of cellular responses. In order to have safe NPs for biomedical applications, the current biocompatibility researches are particularly focused on the severe toxic mechanisms which cause cells death. These mechanisms are apoptosis, autophagy and necrosis, which can also be intricately linked with the cell-life cycle, as there are various check-points and controls in a cell's life cycle to ensure appropriate division processes. Mechanisms by which toxicants induce cell death by necrosis and apoptosis have been the focus of many biomedical disciplines because it helps us understand toxicity but also provides opportunities for drugs to impact on dysregulation of the cell cycle in diseases such as cancer. Among various types of NPs, the superparamagnetic iron oxide nanoparticles (SPION) are recognized as powerful biocompatible materials for multi-task nanomedicine applications such as drug delivery, magnetic resonance imaging, cell/protein separation, hyperthermia and transfection. This chapter presents overview of the effect of SPION on the cell life cycle.

Keywords: Superparamagnetic iron oxide nanoparticles, Cell cycle, TUNEL assay, Protein absorption, Polyethylene glycol fumarate, Polyvinyl alcohol, Propidium iodide, Phosphate buffer saline, Fetal bovine serum, MTT assay, Derivative study.

CELL LIFE CYCLE

The cell life cycle corresponds to a series of events which lead the cell to its division, duplication, and death [1-5]. Cell-life phases are divided into three main parts including G_1, S, and G_2. In the first gap phase (G_1), the cell grows and produces enzymes that are necessary for cell division. In the synthesis phase (S), the DNA is replicated. In the second gap phase (G_2), the cell continues to grow and the cell is carrying out processes necessary for mitosis (M). In both the G_1 and G_2 phases, there are checkpoints that ensure appropriate criteria are met for cycle progression. The effect of NPs on cells depends on their physiochemical properties such as size and distribution, shape, and charge [6]. One adverse effect of certain NPs is the induction of oxidative stress in treated-cells, causing the potential for DNA damage as an early effect evidenced in cell cycle progression. DNA damage is divided into reversible and irreversible types. Considering the cells with reversibly damaged DNA, the cells will accumulate in the G_1, or S, and in the G_2/M phases [7]. Cells which carry irreversibly damaged DNA will proceed to apoptosis, giving rise to the formation of fragmented DNA that can be identified in the subG$_1$ phase [8].

The cell cycle is a vital process for removal of the damaged cells (via apoptosis) and the disruption of this regulated process can induce the formation of tumors. More specifically, some genes like the cell cycle inhibitors (*e.g.,* RB, p53) when they mutate, can cause the cell to multiply uncontrollably, forming a tumor. Although the duration of cell cycle in tumor cells is equal to or longer than that of normal cell cycle, the

Address correspondence to Morteza Mahmoudi: National Cell Bank, Pasteur Institute of Iran, 69 Pasteur Ave. Kargar Ave., Tehran, Iran; Tel: +989125791557; E-mail: mahmoudi@biospion.com

Haseeb Ahmad Khan and Ibrahim Abdulwahid Arif (Eds)

proportion of cells that are in active cell division (versus quiescent cells in G_0 phase) in tumors is much higher than that in healthy tissue. Thus there is a net increase in cell number as the number of cells that die by apoptosis or senescence remains the same. The cells which are actively undergoing cell cycle transition are targeted in cancer therapy as the DNA is relatively exposed during cell division and hence susceptible to damage by drugs or radiation. This physiology is exploited in cancer treatment by a process known as debulking, whereby a significant mass of the tumor is removed which pushes a number of the remaining tumor cells from G_0 to G_1 phase.

SPION

SPION are classified as inorganic-based NPs having an iron oxide core coated by both inorganic and organic materials. There are two types of iron oxides including magnetite (Fe_3O_4) and maghemite (γ-Fe_2O_3), however the magnetite has attracted scientists due to its greater biocompatibility in comparison to maghemite [9, 10]. The favorable inorganic coatings are limited to silica and gold, however there are wide range of organic coatings such as polymers (*e.g.,* polyethylene glycol, polyethylene glycol fumarate (PEGF), and polyvinyl alcohol), acrylates, phospholipids, fatty acids, polysaccharides, and peptides [11]. In comparison with other NPs, SPION have the capability to target a desired site or to heat in the presence of an externally applied AC magnetic field, due to their inducible magnetization. More specifically, SPION have been recognized as a very promising kind of NPs not only due to their very good biocompatibility [11-17], but also due to their diversity of potential applications which can significantly increase patient compliance [18-20].

The SPION have been extensively employed for both *in vitro* and *in vivo* biomedical applications such as magnetic resonance imaging (MRI) contrast enhancement [21, 22], tissue specific release of therapeutic agents [23], hyperthermia, transfection, cell/biomolecules separation, and targeted drug delivery [24]. Many SPION such as Feridex, Endorem or Combidex are commercial and have the FDA approval for MR imaging [25, 26]. The current approaches in SPION are focused on their usage in *'theragnositc'* (*i.e.,* therapeutic and diagnostic) applications.

EFFECT OF SPION ON CELL LIFE CYCLE

The effects of different SPION on the cell-life cycle of various cells are summarized in Table **1**. Preliminary SPION formulations have shown to induce both reversible and irreversible DNA damage. For example, the toxic effects of bare SPION, with both magnetite and maghemite structures, on the A549

Table 1: Effect of SPION on the cell-life assay.

Coating	Size (nm)	Cell Type	Exposure		Phase Arrest	Phase Enhanced	Remark	Refs.
			Conc.	Time (h)				
Polyvinyl alcohol	48	Mouse tissue connective	80 mM	72	None	G_2/M	Surface passivated nanoparticles were used.	[7]
None	4.5	Mouse tissue connective	80 mM	72	None	Sub G_0G_1 and G_2/M	Surface passivated nanoparticles were used.	[7]
Polyvinyl alcohol	12	Mouse tissue connective	200-400 mM	72	None	None	Surface Active nanoparticles were used.	[13]
None	4	Mouse tissue connective	200-400 mM	72	G_0G_1	Sub G_0G_1	Surface Active nanoparticles were used.	[12, 13]
Carboxy-dextran	45-60	human mesenchy-mal stem cells	300 µg/ml	1	None	S and G_2/M	SPION-promoted cell growth is due to its ability to diminish intracellular H_2O_2 through intrinsic peroxidase-like activity.	[35]

human lung epithelial cell line were probed. The abilities of these magnetic nanoparticles to cause DNA damage and oxidative lesions have been evaluated using the comet assay [27]. The intracellular production of reactive oxygen species (ROS) was also measured by the oxidation sensitive fluoroprobe 2',7'-dichlorofluorescin diacetate and the observed toxicity ranged from none to low. Neither DNA damage nor intracellular ROS toxic effects in human lung cells were seen from interaction with magnetite nanoparticles at concentrations of 20-40 µg/mL. However, low quantities of oxidative DNA lesions were observed. There are several methods to track the effects of the DNA damage on the cell-life cycle phases such as the TUNEL (terminal deoxynucleotidyl transferase-mediated dUTP nick end-labeling) assays, which will be described later in this chapter.

APOPTOSIS MEASUREMENT

The apoptosis phenomenon occurs due to the irreversible DNA damages. A ubiquitous feature of the apoptosis phenomenon is the breakup of chromatin, which happens during the exposure of numerous 3' OH DNA ends. By analyzing the DNA of cells which are undergoing apoptosis, using gel electrophoresis, a unique ladder-like appearance of DNA pieces with discrete molecular weights is observed. Hence, a reliable and rapid method for apoptosis evaluation is to compare the mobility of DNA extracted from control and apoptotic cells, for instance comparing DNA mobility of untreated Jurkat cells to the mobility of DNA of camptothecin-induced Jurkat cells [28, 29]. To determine apoptosis due to the exposure of cells to the SPION, there are several commercialized kits such as Apoptosis APO-BRDUTM kit (Sigma-Aldrich, Inc.) which needs dual color flow cytometry method for its evaluation.

Typically, BRDUTM kit provides a simple process of assessing apoptosis; however, the use of this kit requires that the cells are lysed. The appearance of the 3' OH ends can also be quantified as a measurement of apoptosis in whole cells by an alternative method which does not require cell lysis. An alternative method in mixed cell populations is called the TUNEL assay, also known as the bromodeoxyuridine terminal deoxynucleotidyl transferase assay. For instance, the L929 mouse fibroblasts connective tissue cells were treated with both bare and polyvinyl alcohol (PVA)-coated SPION and their apoptotic effects were tracked with the TUNEL assay [12]. In order to prepare both control and treated-cells for flow cytometry evaluations, the predetermined cells were fixed with paraformaldehyde in PBS, followed by ethanol fixation. Consequently, the cells were washed and reacted with the TdT enzyme (terminal deoxynucleotidyl transferase) and Br-dUTP (bromodeoxyuridine triphosphate) in buffered solution at 37°C for 60 min. In this case, bromodeoxyuridine was covalently incorporated into the 3' DNA ends during this incubation. Cells should be then thoroughly rinsed and incubated with a FITC (fluorescein isothiocyanate) labeled antibody directed to bromodeoxyuridine for about 30 min. After washing away unbound antibody, immunostaining with the FITC labeled anti-bromodeoxyuridine antibody allowed to determine the number of free 3' ends. The RNA of the cells was then digested and the total DNA stained by incubation with a solution containing RNase A plus propidium iodide. Staining of cells with propidium iodide allows normalizing FITC staining to the total amount of DNA in the cells. Finally, the stained cells were analyzed by flow cytometry with an argon laser emitting at 488 nm. FITC fluorescence was observed at 520 nm and propidium iodide simultaneously at 623 nm. The results shown in Fig. **1** indicated that the SPION treated-cells did not show apoptosis at the examined SPION content up to concentrations of 200 mM.

CELL CYCLE ASSAY

Cell cycle assay could be evaluated by staining of the DNA with the suitable fluorescence dyes, such as propidium iodide (PI), followed by flow cytometric measurement of the fluorescence. Typically, the cells were cultured and then treated with the NPs for the desired time. Since the damaged cells may leave their attached places and be suspended in medium, the medium should be stored after removal.

Then, the remaining adhesive cells could be detached from the flask *via* trypsin treatment and harvested using the stored medium followed by centrifugation at about 280 G. The collected cells were washed thoroughly with phosphate buffer saline (PBS) followed by transferring of cells into the tubes containing 70% ethanol for fixation and stored at -20°C. Prior to the flow cytometric analysis, the ethanol-suspended cells were centrifuged at 200g for about 5 min and the supernatants were decanted comprehensively. The collected cells were washed with PBS and then suspended in 1 ml PI/Triton X-100 staining solution with RNase A at 37°C for 15 min or at ambient temperature for 30 min. The stained cells are then ready for evaluations by flow cytometry at an excitation wavelength of 488 nm (for PI) and emission wavelength of 610 nm.

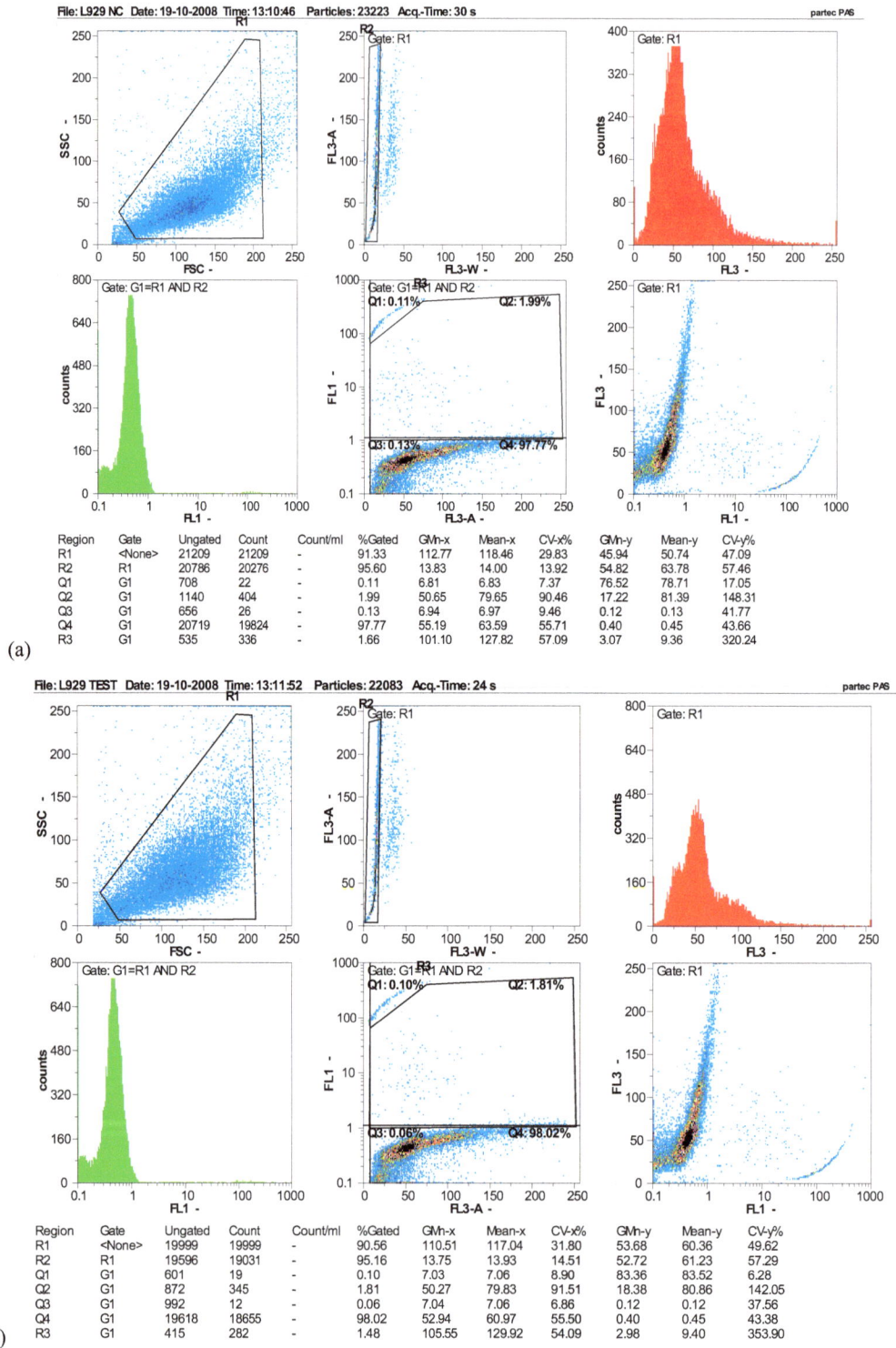

Figure 1: Flow cytometry results for L929 Cells (a) with no SPION added and (b) with SPION (iron concentration of 200 mM) added; with permission from reference [12].

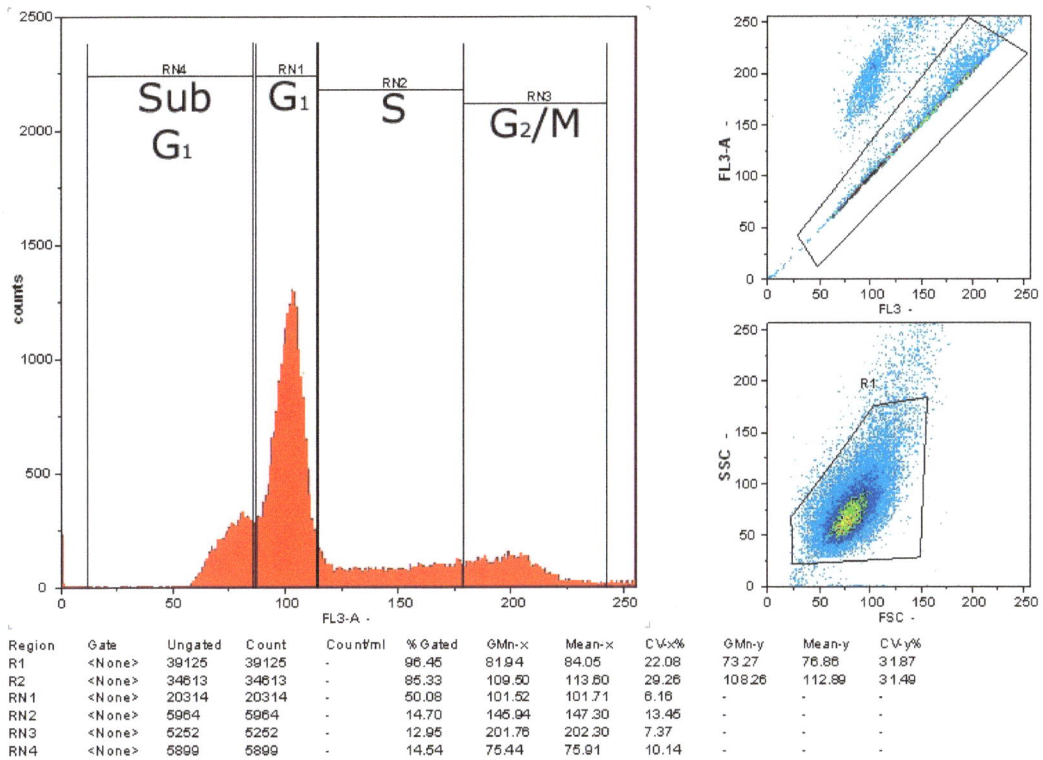

Region	Gate	Ungated	Count	Count/ml	% Gated	GMn-x	Mean-x	CV-x%	GMn-y	Mean-y	CV-y%
R1	<None>	39125	39125	-	96.45	81.94	84.05	22.08	73.27	76.86	31.87
R2	<None>	34613	34613	-	85.33	109.50	113.50	29.26	108.26	112.99	31.49
RN1	<None>	20314	20314	-	50.08	101.52	101.71	6.18	-	-	-
RN2	<None>	5964	5964	-	14.70	145.94	147.30	13.45	-	-	-
RN3	<None>	5252	5252	-	12.95	201.76	202.30	7.37	-	-	-
RN4	<None>	5899	5899	-	14.54	75.44	75.91	10.14	-	-	-

(a)

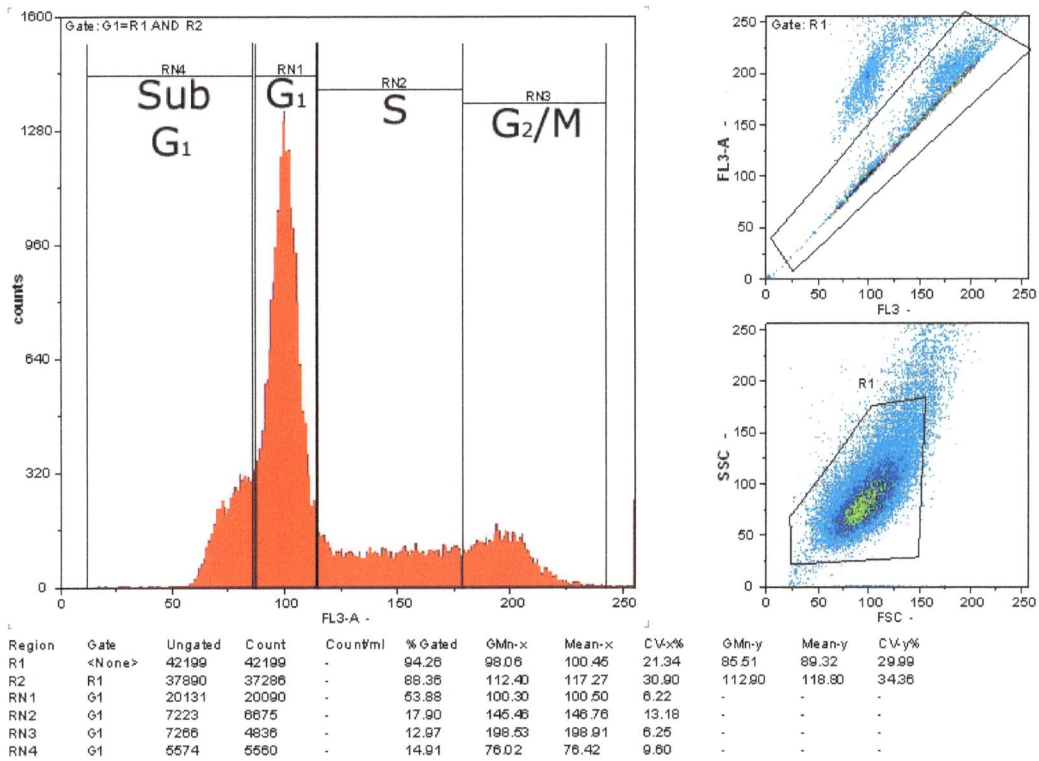

Region	Gate	Ungated	Count	Count/ml	% Gated	GMn-x	Mean-x	CV-x%	GMn-y	Mean-y	CV-y%
R1	<None>	42199	42199	-	94.26	98.06	100.45	21.34	85.51	89.32	29.99
R2	R1	37890	37286	-	88.36	112.40	117.27	30.90	112.90	118.80	34.36
RN1	G1	20131	20090	-	53.88	100.30	100.50	6.22	-	-	-
RN2	G1	7223	6875	-	17.90	145.46	146.76	13.18	-	-	-
RN3	G1	7266	4836	-	12.97	198.53	198.91	6.25	-	-	-
RN4	G1	5574	5560	-	14.91	76.02	76.42	9.60	-	-	-

(b)

Figure 2: Cell cycle assay results for (a) control and (b) coated SPION (200 mM) treated cells; with permission from reference [13].

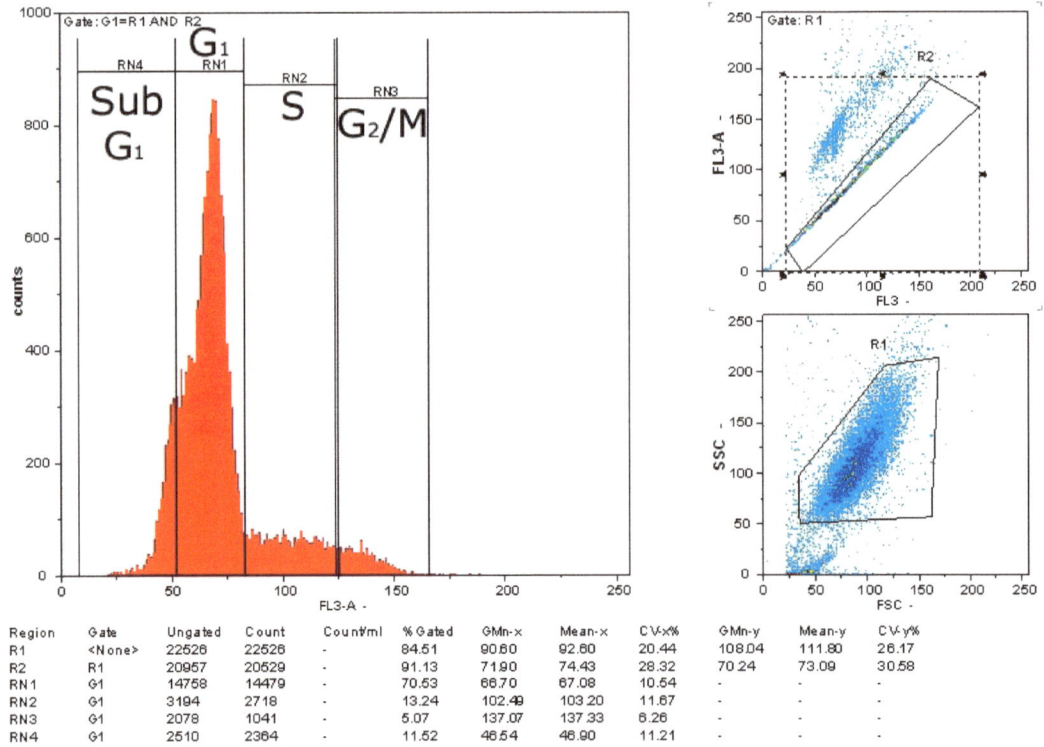

Region	Gate	Ungated	Count	Count/ml	% Gated	GMn-x	Mean-x	CV-x%	GMn-y	Mean-y	CV-y%
R1	<None>	22526	22526	-	84.51	90.60	92.60	20.44	108.04	111.80	26.17
R2	R1	20957	20529	-	91.13	71.90	74.43	28.32	70.24	73.09	30.58
RN1	G1	14758	14479	-	70.53	66.70	67.08	10.54	-	-	-
RN2	G1	3194	2718	-	13.24	102.49	103.20	11.67	-	-	-
RN3	G1	2078	1041	-	5.07	137.07	137.33	6.26	-	-	-
RN4	G1	2510	2364	-	11.52	46.54	46.90	11.21	-	-	-

(a)

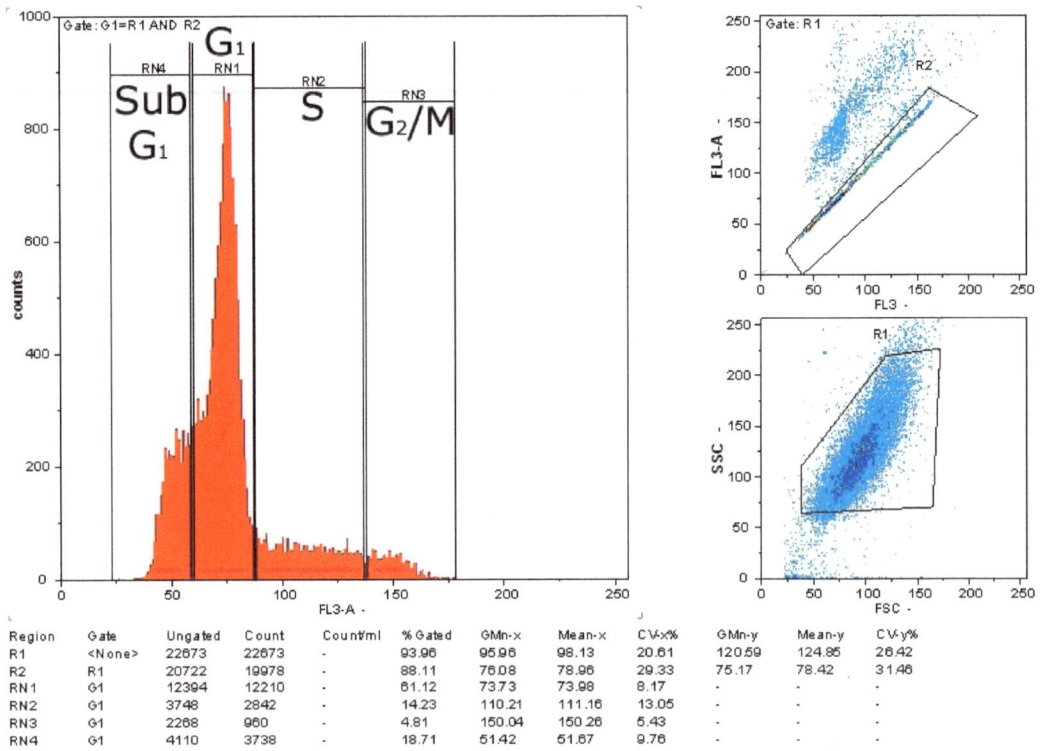

Region	Gate	Ungated	Count	Count/ml	% Gated	GMn-x	Mean-x	CV-x%	GMn-y	Mean-y	CV-y%
R1	<None>	22673	22673	-	93.96	95.96	98.13	20.61	120.59	124.85	26.42
R2	R1	20722	19978	-	88.11	76.08	78.96	29.33	75.17	78.42	31.46
RN1	G1	12394	12210	-	61.12	73.73	73.98	8.17	-	-	-
RN2	G1	3748	2842	-	14.23	110.21	111.16	13.05	-	-	-
RN3	G1	2268	960	-	4.81	150.04	150.26	5.43	-	-	-
RN4	G1	4110	3738	-	18.71	51.42	51.67	9.76	-	-	-

(b)

Figure 3: Cell cycle assay results for (a) control and (b) coated SPION (400 mM) treated cells; with permission from reference [13].

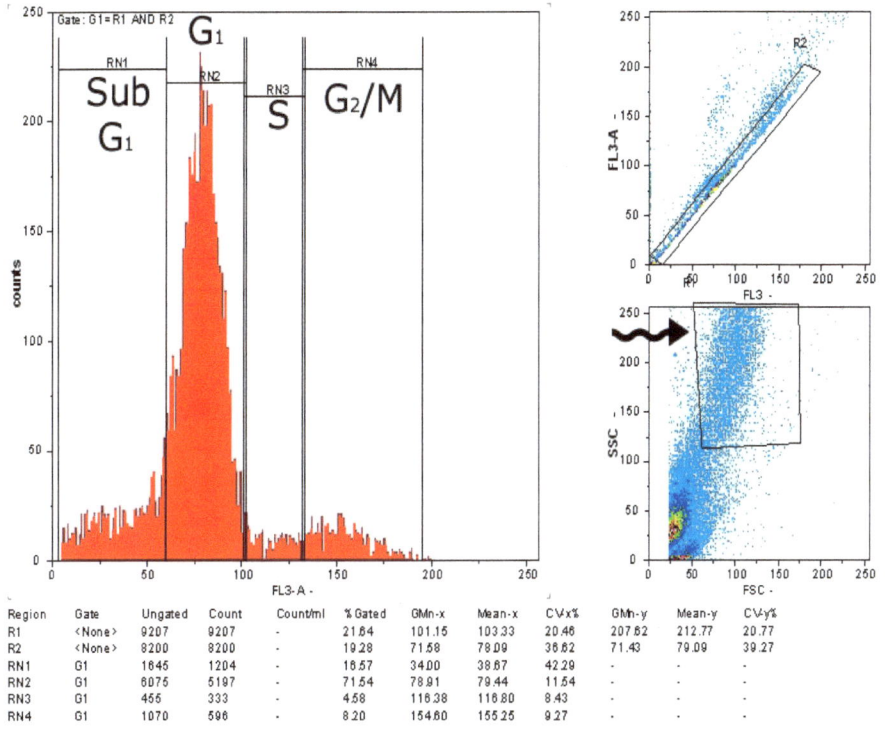

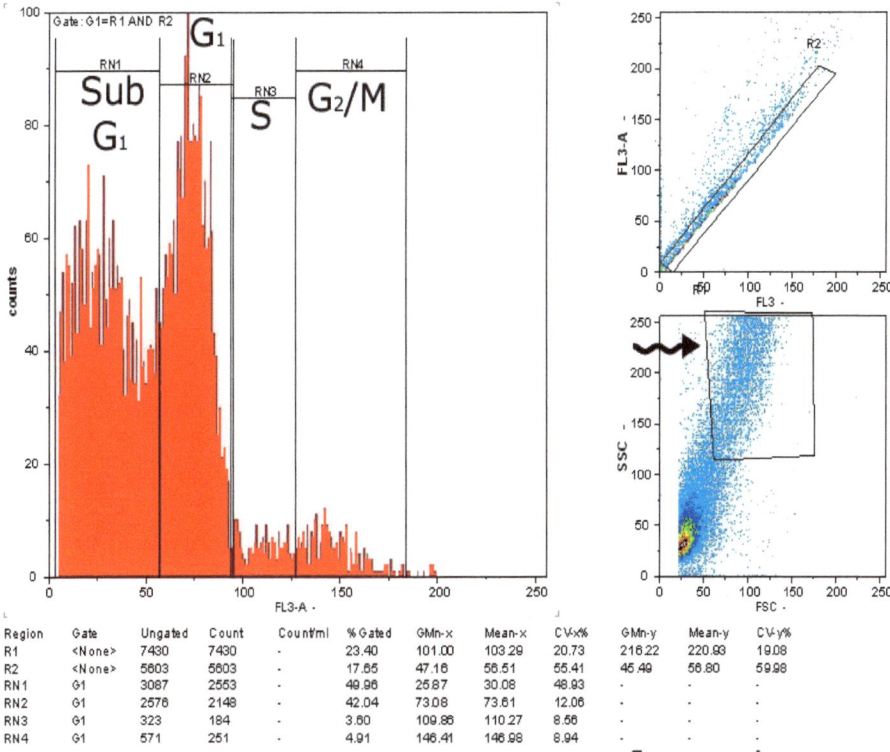

Figure 4: Cell cycle assay results for treated cells to uncoated SPION with concentration of (a) 200 and (b) 400 mM; with permission from reference [13].

In order to track the effects of both bare and PVA coated SPION on the L929 cells, this method has been applied. Cell cycle assessment was carried out by staining of the DNA with PI followed by flow cytometric measurement [13]. Approximately 10^6 L929 cells were cultured and treated with SPION with concentrations of 100, 200 and 400 mM of iron for 72 h. The effects of SPION treated cells were probed in each phase of the cell cycle and compared with control cells. According to the obtained results, both bare and PVA coated SPION with the iron concentration of 100 and 200 mM have no detectable effect on the cell life cycle phases and were similar to the control cells (see Figs. **2** and **4**).

The same proportion of the cell population in subG$_1$ phase in control cells and SPION treated cells confirmed the absence of apoptosis. As the concentration of coated-SPION increased with an iron concentration of 400 mM, a negligible amount of apoptosis was observed in the assay (see Fig. **3**). In contrast, for the bare SPION treated cells, a significant increase in the proportion of apoptotic cells (ΔSubG$_1$/SubG$_1$(control) =0.62) was observed (see Fig. **4**) due to the irreversible DNA damage. In addition to apoptosis, the arrest in the G$_0$G$_1$ phase was detected for the bare SPION treated cells at an iron concentration of 400 mM. Furthermore, due to the higher surface activity of the bare SPION in comparison with the PVA coated particles, the granularity of the bare SPION treated-cells was increased, which is clearly shown in Fig. **4b**.

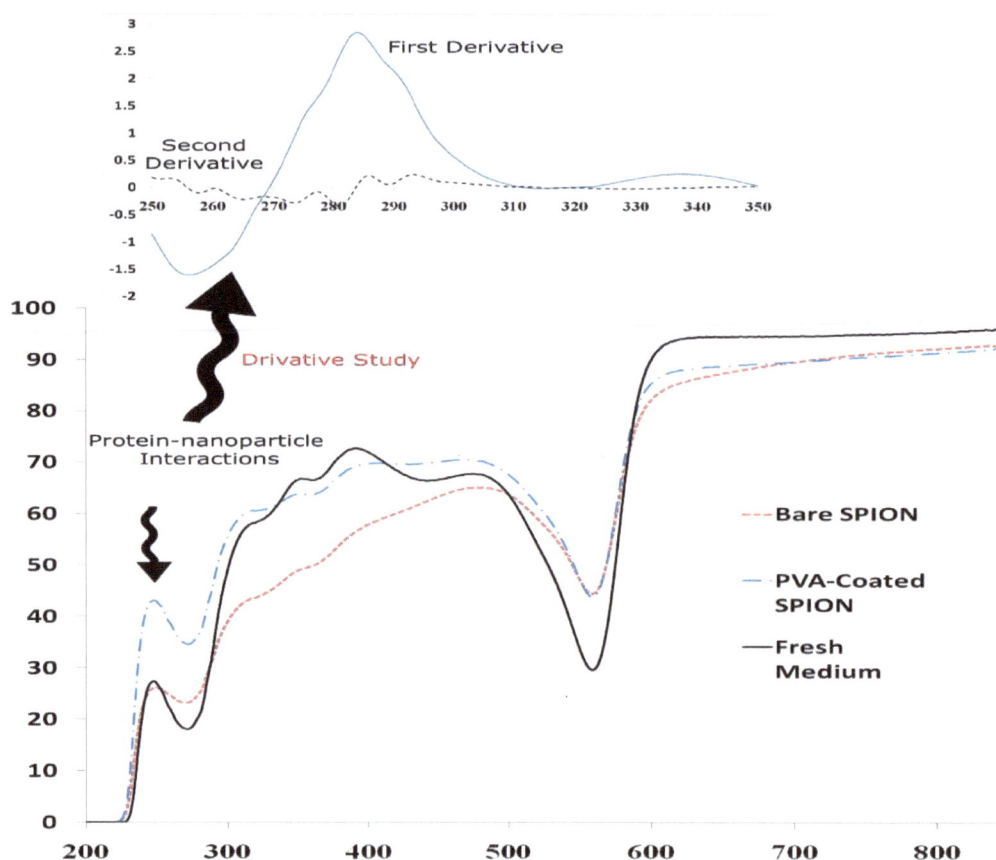

Figure 5: UV/Vis spectra of fresh cell medium, the extract medium of uncoated and coated SPION (DI water defined as reference in UV/Vis spectrum). Upper graph: first (solid line) and second (dash-line) order derivatives of coated SPION (with iron concentration of 400 mM); with permission from reference [13].

PROTEIN-SPION INTERACTIONS AND THEIR EFFECTS ON CELL CYCLE ASSAY

It is now well-recognized that the NPs can interact with proteins on their entrance into the biological environment [30-34]. The outcome of these interactions may cause the malfunction of the crucial proteins that could have a significant effect on the cell life cycle. *In vitro* investigations have shown that SPION can interaction with cell medium containing fetal bovine serum (FBS) whereby the particles demonstrate adsorption of FBS proteins to the surface of the SPION. More specifically, the UV/Vis spectroscopy of both fresh cell medium and extracted medium after interactions with bare and coated (PVA and PEGF) SPION confirmed not only protein absorption but also pH changes (at wavelength of 560 nm in the spectra) of the cell culture which can cause severe errors in toxicity evaluation methods (*e.g.,* MTT (3-(4,5-dimethylthiazol-2-yl)-2,5-diphenyltetrazolium bromide) assay) [13, 15, 16]. Fig. **5** shows the effect of uncoated and PVA-coated SPION on the cell medium. In addition to PVA, the effect of PEGF coated SPION on the cell medium were probed and similar results were obtained (Fig. **6**).

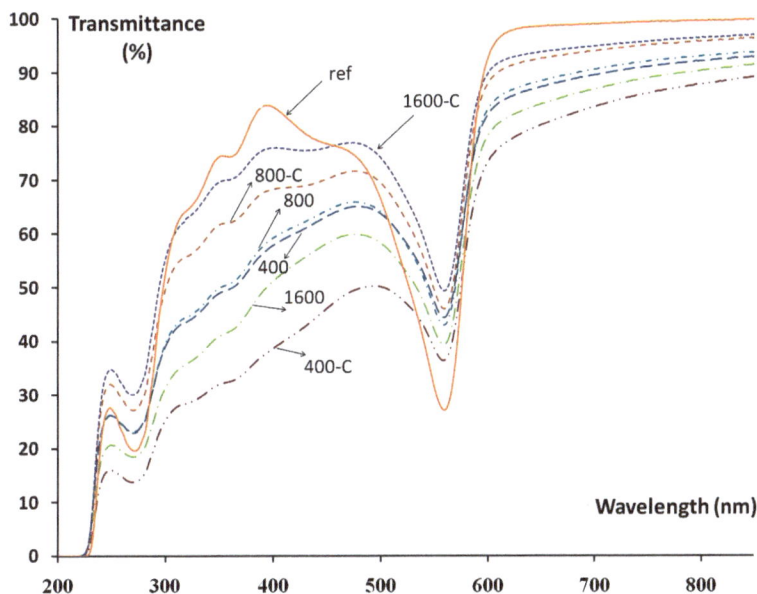

Figure 6: UV/Vis spectra of 'pure' cell medium (ref), the extract of coated (C) and bare SPION with de-ionized (DI) water as the reference in UV/Vis spectrum. (The numbers are concentrations of iron in mM, C refers to coated nanoparticles); with permission from reference [15].

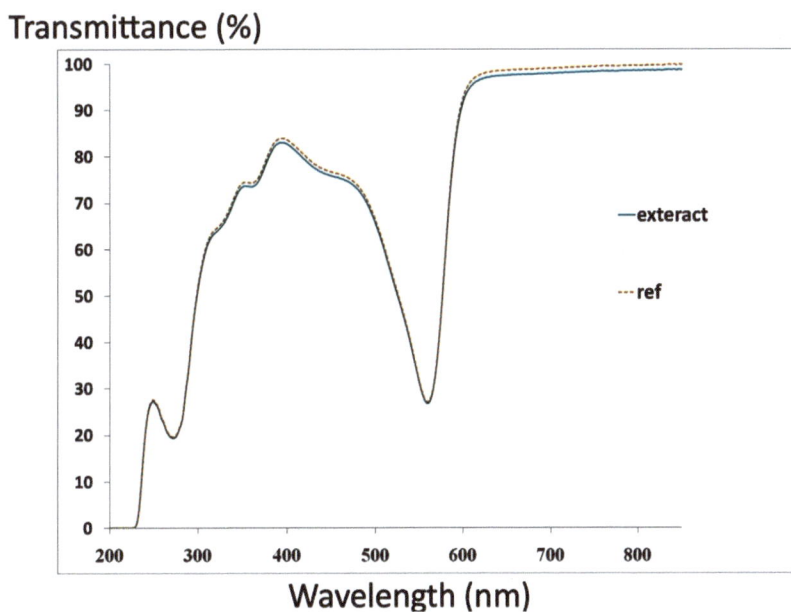

Figure 7: UV/Vis spectra of pure cell medium (ref) and the extract of surface saturated SPION; with permission from reference [15].

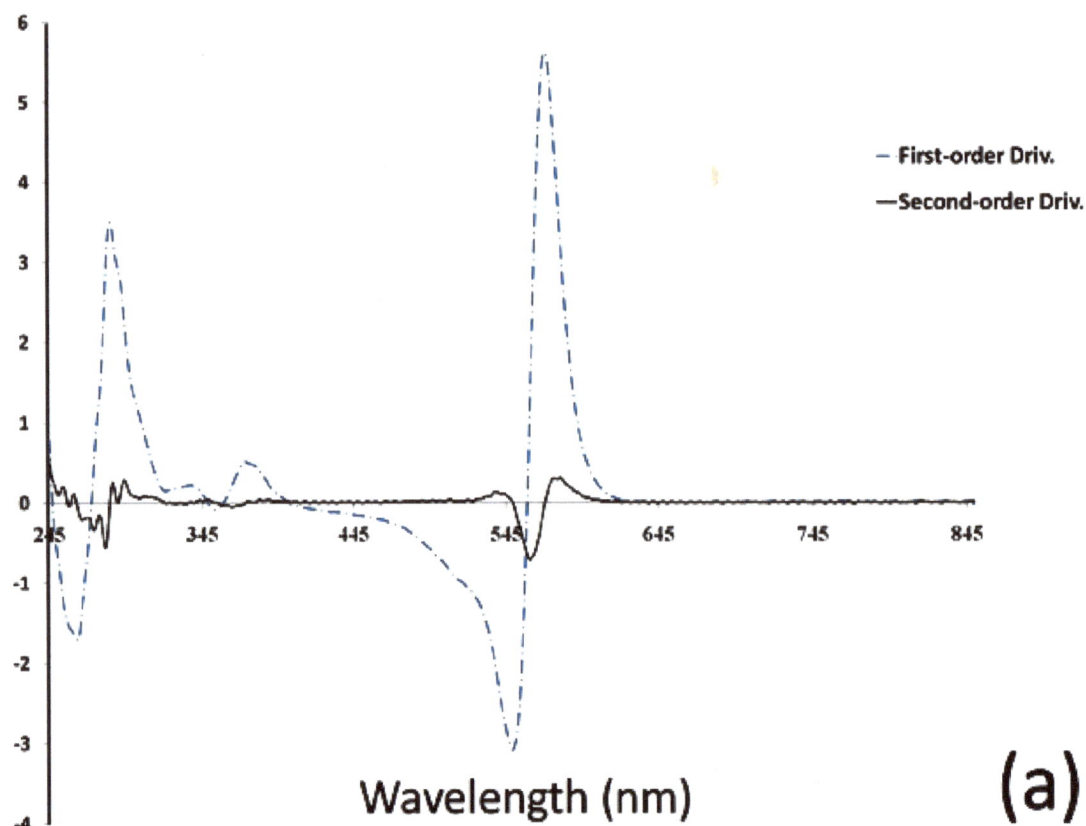

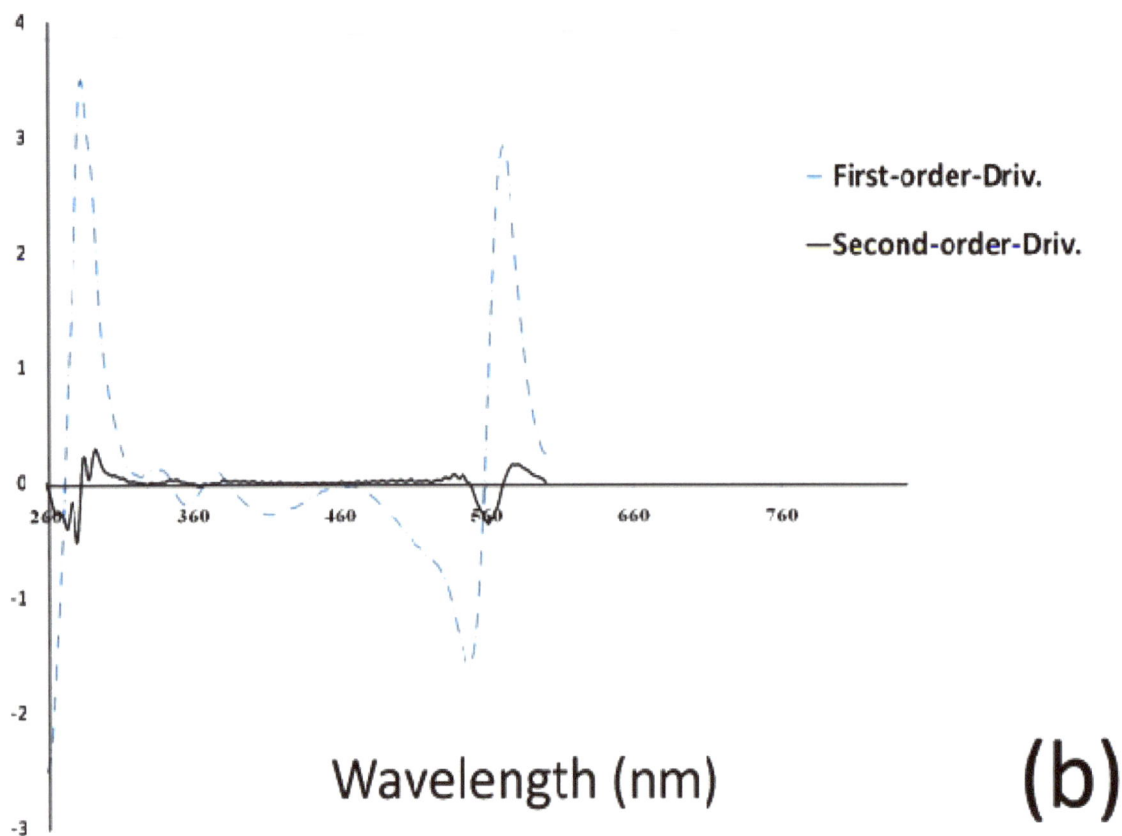

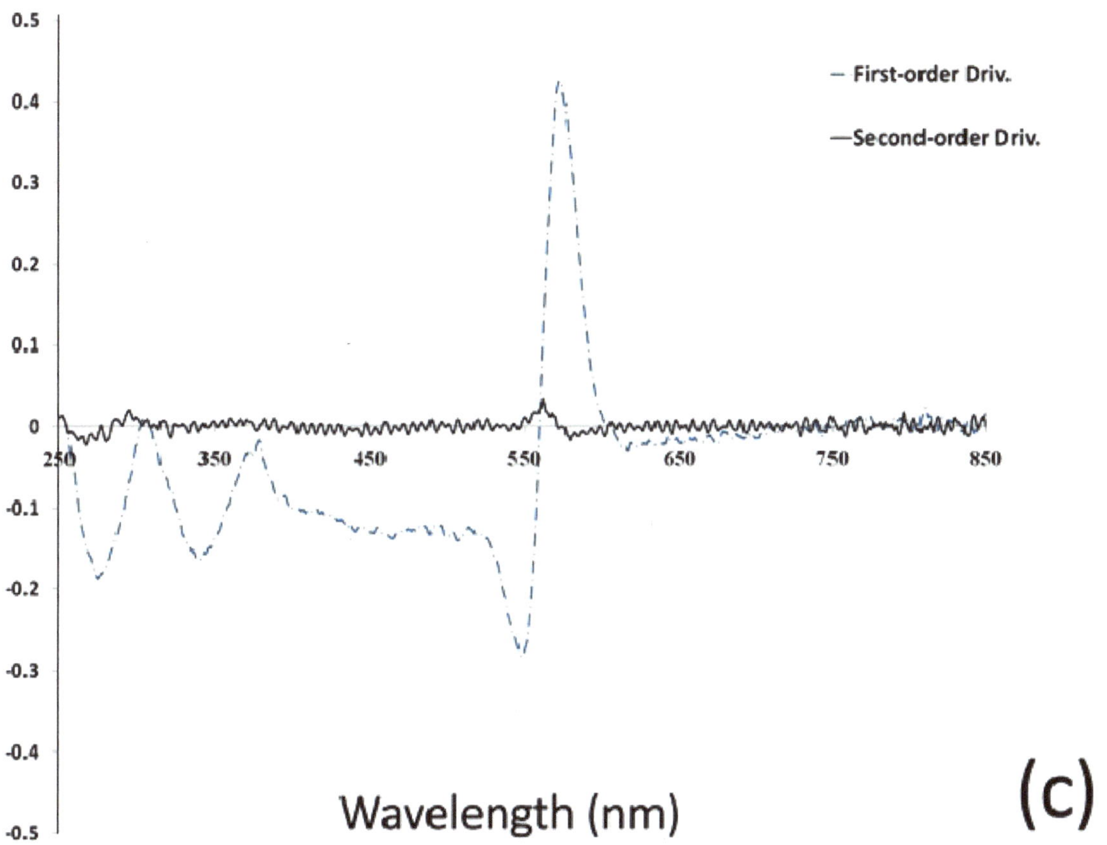

Figure 8: First (dash-line) and second (solid line) order derivatives of (a) 1600B, (b) 1600C, and (c) 1600CM (modified sample); with permission from reference [15].

To overcome this problem, a modified method has been proposed using the following steps: (1) introduction of the nanoparticles to the cell medium, (2) incubation of the solution for 24 hours in order to create a hard protein corona, (3) replacing the medium with a fresh one, and (4) application of the surface saturated SPION to the assays [13, 15, 16]. By applying this protocol, the obtained SPION did not change the spectra of the cell medium (see Fig. **7**), resulting a more appropriate toxicity evaluations of the NPs.

In order to prove that the adsorption of several proteins to the surface of SPION occurs, the derivative spectroscopy method was employed [13, 15, 16]. Both first and second order derivatives have been probed for detection of multi-component changes in cell culture medium following interaction with SPION. The results are shown in Fig. **8**. Multi-protein absorbance is detected *via* the second derivative curves between the wavelengths of 250-300 nm in both bare and PEGF-coated SPION (Figs. **8a** and **b**). Fig. **8c** shows the derivative curves for the modified samples, denoting negligible absorbance at aforementioned wavelengths. By applying this new protocol to the L929 cell lines, the observed irreversible DNA damages are decreased leading to the reduction of apoptosis of the cells at the same concentrations [7].

LIST OF ABBREVIATIONS

NPs: Nanoparticles; SPION: Superparamagnetic Iron Oxide Nanoparticles; MRI: Magnetic Resonance Imaging; ROS: Reactive Oxygen Species; TUNEL: Terminal deoxynucleotidyl transferase-mediated dUTP nick end-labeling; PVA: PolyVinyl Alcohol; PI: Propidium Iodide; PBS: Phosphate Buffer Saline; FBS: Fetal Bovine Serum; PEGF: PolyEthylene Glycol Fumarate; MTT: 3-(4,5-dimethylthiazol-2-yl)-2,5-diphenyltetrazolium bromide

REFERENCES

[1]   Smith JA, Martin L. Do cells cycle? Proc Natl Acad Sci USA 1973; 70 (4): 1263-1270.

[2]   Morgan DO. The Cell Cycle: Principles of Control. London, New Science Press in association with Oxford University Press 2007.

[3]   Lewin B, Cassimeris L, Lingappa VR. Cells, Jones & Bartlett Publishers 2007.

[4]   Alberts B, Johnson A, Lewis J, Raff M, Roberts K, Walter P. Molecular Biology of the Cell (5th ed.). New York: Garland Science 2008.

[5]   Alberts B, Bray D, Johnson A, *et al.* Essential Cell Biology: An Introduction to the Molecular Biology of the Cell Garland Publishing, New York 1997.

[6]   Nel AE, Madler L, Velegol D, *et al.* Understanding biophysicochemical interactions at the nano-bio interface. Nat Mater 2009; 8: 543-557.

[7]   Mahmoudi M, Simchi A, Vali H, *et al.* Cytotoxicity and cell cycle effects of bare and polyvinyl alcohol coated iron oxide nanoparticles in mouse fibroblasts Adv Eng Mater 2009; 11 (12): B243-B250.

[8]   Ishikawa K, Ishii H, Saito T. DNA Damage-Dependent Cell Cycle Checkpoints and Genomic Stability. DNA Cell Biol 2006; 25: 406-411.

[9]   Gupta AK, Gupta M, Cytotoxicity suppression and cellular uptake enhancement of surface modified magnetic nanoparticles. Biomaterials 2005; 26: 1565-1573.

[10] Gupta AK, Naregalkar RR, Vaidya VD, Gupta M. Recent advances on surface engineering of magnetic iron oxide nanoparticles and their biomedical applications. Nanomedicine 2007; 2 (1): 23-39.

[11] Mahmoudi M, Simchi A, Imani M. Recent advances in surface engineering of superparamagnetic iron oxide nanoparticles for biomedical applications. J Iranian Chem Soc 2010; 7 (3): 1-27.

[12] Mahmoudi M, Shokrgozar MA, Simchi A, *et al.* Multiphysics Flow Modeling and *in vitro* Toxicity of Iron Oxide Nanoparticles Coated with Poly(vinyl alcohol). J Phys Chem C 2009; 113 (6): 2322-2331.

[13] Mahmoudi M, Simchi A, Imani M. Cytotoxicity of Uncoated and Polyvinyl Alcohol Coated Superparamagnetic Iron Oxide Nanoparticles. J Phys Chem C 2009; 113 (22): 9573-9580.

[14] Mahmoudi M, Simchi A, Imani M, Hafeli UO. Superparamagnetic Iron Oxide Nanoparticles with Rigid Cross-linked Polyethylene Glycol Fumarate Coating for Application in Imaging and Drug Delivery. J Phys Chem C 2009; 113 (19): 8124-8131.

[15] Mahmoudi M, Simchi A, Imani M, Milani AS, Stroeve P. An *in vitro* study of bare and poly(ethylene glycol)-co-fumarate-coated superparamagnetic iron oxide nanoparticles: a new toxicity identification procedure. Nanotechnology 2009; 20 (22): 225104.

[16] Mahmoudi M, Simchi A, Imani M, *et al.* A new approach for the *in vitro* identification of the cytotoxicity of superparamagnetic iron oxide nanoparticles. Colloid Surf B 2010; 75: 300-309.

[17] Mahmoudi M, Simchi A, Milani AS, Stroeve P. Cell toxicity of superparamagnetic iron oxide nanoparticles. J Colloid Interface Sci 2009; 336: 510-518.

[18] Mahmoudi M, Hosseinkhani H, Hosseinkhani M, *et al.* MRI tracking of stem cells *in vivo* using iron oxide nanoparticles as a tool for the advancement of clinical regenerative medicine. Chem Rev 2011; 111 (2): 253-280.

[19] Mahmoudi M, Milani AS, Stroeve P. Surface Architecture of Superparamagnetic Iron Oxide Nanoparticles for Application in Drug Delivery and Their Biological Response: A Review. Int J Biomedical Nanoscience and Nanotechnology 2010; 1(2/3/4/): 164-201.

[20] Mahmoudi M, Sant S, Wang B, *et al.* Superparamagnetic Iron Oxide Nanoparticles (SPION): Development, surface modification and applications in chemotherapy. Adv Drug Delivery Rev 2011; 63(1-2): 24-46.

[21] Cunningham CH, Arai T, Yang PC, *et al.* Positive contrast magnetic resonance imaging of cells labeled with magnetic nanoparticles. Magn Res Med 2005; 53 (5): 999-1005.

[22] Anderson SA, Rader RK, Westlin WF, *et al.* Magnetic resonance contrast enhancement of neovasculature with alpha(v)beta(3)-targeted nanoparticles. Magn Res Med 2000; 44 (3): 433-439.

[23] Polyak B, Friedman G. Magnetic targeting for site-specific drug delivery: applications and clinical potential. Exp Opin Drug Delivery 2009; 6 (1): 53-70.

[24] Jalilian AR, Panahifar A, Mahmoudi M, Akhlaghi M, Simchi A. Preparation and biological evaluation of [67Ga]-labeled- superparamagnetic nanoparticles in normal rats. Radiochim Acta 2009; 97 (1): 51-56.

[25] Bartolozzi C, Lencioni R, Donati F, Cioni D. Abdominal MR: Liver and pancreas. Europ Radiol 1999; 9 (8): 1496-1512.

[26]   Meng J, Fan J, Galiana G, *et al.* LHRH-functionalized superparamagnetic iron oxide nanoparticles for breast cancer targeting and contrast enhancement in MRI. Mater Sci Eng C 2009; 29 (4): 1467-1479.

[27]   Karlsson HL, Cronholm P, Gustafsson J, Moller L. Copper oxide nanoparticles are highly toxic: A comparison between metal oxide nanoparticles and carbon nanotubes. Chem Res Toxicol 2008; 21: 1726-1732.

[28]   Saleh OA, Blalock WL, Burrows C, *et al.* Enhanced ability of the progenipoietin-1 to suppress apoptosis in human hematopoietic cells. Int J Mol Med 2002; 10 (4): 385-394.

[29]   Reinhold WC, Kouros-Mehr H, Kohn KW, *et al.* Apoptotic susceptibility of cancer cells selected for camptothecin resistance: Gene expression profiling, functional analysis, and molecular interaction mapping. Cancer Res 2003; 63 (5): 1000-1011.

[30]   Cedervall T, Lynch I, Foy M, *et al.* Detailed Identification of Plasma Proteins Adsorbed on Copolymer Nanoparticles. Ang Chem Int Ed 2007; 46: 5754 -5756.

[31]   Cedervall T, Lynch I, Lindman S, *et al.* Understanding the nanoparticle-protein corona using methods to quantify exchange rates and affinities of proteins for nanoparticles. Proc Nat Acad Sci USA 2007; 104 (7): 2050-2055.

[32]   Lundqvist M, Stigler J, Elia G, Lynch I, Cedervall T, Dawson KA. Nanoparticle size and surface properties determine the protein corona with possible implications for biological impacts. Proc Nat Acad Sci USA 2008; 105 (38): 14265-14270.

[33]   Lynch I, Cedervall T, Lundqvist M, Cabaleiro-Lago C, Linse S, Dawson KA, The nanoparticle-protein complex as a biological entity; a complex fluids and surface science challenge for the 21st century. Adv Colloid Interface Sci 2007; 134-135: 167-174.

[34]   Lynch I, Dawson KA, Linse S. Detecting Cryptic Epitopes Created by Nanoparticles. Science 2006; 327: 14.

[35]   Huang DM, Hsiao JK, Chen YC, *et al.* The promotion of human mesenchymal stem cell proliferation by superparamagnetic iron oxide nanoparticles. Biomaterials 2009; 30 (22): 3645-3651.

Safety of Magnetic Iron Oxide-Coated Nanoparticles in Clinical Diagnostics and Therapy

Safety of Magnetic Iron Oxide-Coated Nanoparticles in Clinical Diagnostics and Therapy

Ângela Leao Andrade[1], Rosana Zacarias Domingues[2], José Domingos Fabris[2,3] and Alfredo Miranda Goes[4]

[1]*Department of Chemistry, ICEB, Federal University of Ouro Preto, 35400-000 Ouro Preto, Minas Gerais, Brazil;* [2]*Department of Chemistry, ICEx, UFMG, Campus - Pampulha, 31270-901 Belo Horizonte, Minas Gerais, Brazil;* [3]*Department of Chemistry, Federal University of Jequitinhonha and Mucuri Valleys, 39100-000 Diamantina, Minas Gerais, Brazil and* [4]*Department of Biochemistry and Immunology, ICB, UFMG, Campus - Pampulha, 31270-901 Belo Horizonte, Minas Gerais, Brazil*

Abstract: Most potential benefits of nanotechnology in industrial processes and in medicine are inimitable. The versatility of magnetic nanoparticles (MNP) is mainly due to their small size-induced properties, which govern their ability to readily respond to an external magnetic field and their achievable functionality as bioactive agents. Both of these two characteristics may be built either individually or be suitably combined. However, any potentially harmful consequence of MNPs to the integrity of healthy human tissues must be safely and seriously taken into account. Scientific researcher, commercial manufacturers and government staffs along with all those expertises concerned with eventual pernicious effects of chemical synthetic products on the natural environment increasingly require futher studies on toxicity effects of nanoparticles, particularly as constituents of those materials more often used in commercial products destined to internal medicine therapy or dignostic procedures, or to other less critical toxic principles for humans, such as in daily used over-skin creams and body cosmetics. Regulatory laws on the use or manipulation of toxic products rely on the accuracy and right choice of test protocols to evaluate their real organic safety. Multidisciplinary studies envolving nanomaterials, toxicitivity as well as biomedical and other disciplins will certainly guide the development of advanced and futurely still more biocompatible materials and devices to be used in medical practices.

Keywords: Magnetic nanoparticles, Iron oxide nanoparticles, Safety, Clinical applications, Diagnostics, Therapy.

INTRODUCTION

Some intrinsic properties of nanoparticles make these compounds promising candidates in both industry and biomedical applications [1-5]. Reducing particle sizes may dramatically change some of their properties, such as electrical conductivity, magnetic characteristics, hardness, active surface area, chemical reactivity, and biological activity, relatively to characteristics of the bulk counterpart of similar materials. Due to these changes, at present, the manufacture and use of nanoparticles in hundreds of commercial products is increasingly a new perspective in technology. Engineered nanoparticles are used in tires, clothes, sunscreens, cosmetics, and electronics, and are being increasingly used in medicine [6]. The magnetic features of nanosized particles have unique advantage in medical diagnostic and therapy. Molecular separation, immunoassay, magnetic resonance imaging (MRI), drug delivery, and hyperthermia are being highly improved with the use of magnetic nanoparticles. Magnetite cationic liposomes, one of the groups of cationic magnetic particles, can be used as carriers to introduce DNA into cells as their positively charged surface associates with the negatively charged DNA. They can also be used as heat mediators in cancer therapy. Magnetic particles conjugated with tumor-specific antibodies have allowed the enhancement of tumor-specific contrast in MRI. In addition, antibody-conjugated magnetic particles were shown to target renal cell carcinoma cells, and are applicable to the hyperthermic treatment of carcinomas.

Address correspondence to Rosana Zacarias Domingues: Universidade Federal de Minas Gerais / Departamento de Química, AV. Antonio Carlos, 6627 – Pampulha, Caixa Postal Química: 702-Belo Horizonte, Cep: 30161-970 Minas Gerais, Brazil; Tel: +55 31 34095770 / 34096383; Fax: +55 3134095700; E-mail: rosanazd@qui.ufmg.br

The use of magnetic particles with their unique features is expected to improve further medical techniques [7]. Since nanomaterials are designed to use optimal amounts of active materials, as therapeutical drugs, catalysts etc, providing even much better local effects, they are also economically and environmentally more favorable alternatives [8]. Even though these features may sometimes be impressive from a material science perspective the possibility of causing toxic effects should not be neglected [9]. For instance, certain nanomaterials, which were designed to release some sort of chemical reactants in the environment, may undergo itself side-reactions yielding non-expected hazardous products [10, 11]; little is known about the toxic effects of these so reactive materials. Non-predictable effects in human health or in the environmental, the increase number of cases due to respiratory or cardiovascular mortality and morbidity, the worsening of asthma symptoms [12-15] associated with exposure to engineered nanomaterials, raise questions about potential risks of non-controlled exposure to nanoparticles [16-20]. Solid nanosized (<100 nm) particles are very easily inhaled, ingested, or absorbed. If these materials are allowed to travel throughout the human body, they may impose a significant risk to health, as their mean physical dimension is comparable to that of typical intra-cellular components and proteins [21]. Nowadays, the over-excitement on the potentiality of nanotechnology is being somehow tempered by concerns that scientists are dabbling too far into the unknown. Media reports questioning the wisdom of industry unleashing possibly uncontrollable substances into the wider environment have contributed to an aura of uncertainty and even fear. Concerned parties have called for further research to amass more data on the likely behavior of nanomaterials throughout their lifetime. Regulatory agencies have begun assessing what, if any, steps may need to be taken to monitor or control nanoengineered products. Activists and ethicists have urged governments to impose a ban on further research until the potential risks are better understood [8]. A number of authors have reviewed about the characterization, preparation, fate, and toxicological information on nanomaterials and proposed research strategies for evaluating the safety on their use [22-28]. Evaluating nanomaterials safety means also to consider their interfacial behavior, including their interaction with proteins, DNA, lipids, membranes, organelles, cells, tissues, and biological fluids [29, 30]. A challenge in evaluating risks associated with the manufacture and use is related to the diversity and complexity of the types of synthetic materials commercially available and new products being developed, as well as what seems to be their limitless potential uses.

A risk assessment is the evaluation of scientific information on the hazardous properties of environmental agents, the dose-response relationship, and the extent of exposure of humans or environmental receptors to those agents. The product of the risk assessment is a statement regarding the probability that humans (populations or individuals) or other environmental receptors so exposed will be harmed and to what degree (risk characterization).

NANOTECHNOLOGY IN MEDICINE

Magnetic nanoparticles (MNPs) are a class of nanoparticles which has less than 100 nm in diameter and are usually constituted of magnetic elements, such as iron, nickel, cobalt or their oxides. Because they are sensible to the action of externally applied magnetic field many important applications arise in medicine and other areas. Despite the pros and cons of using these materials for *in vivo* applications, iron oxide MNPs principally as their stable oxides, magnetite (Fe_3O_4) and maghemite (γ-Fe_2O_3), have been approved for clinical use to date [31]. Both presents chemical stability and biocompatibility [32-36]. Fe_3O_4 nanoparticles are normally obtained by various chemical-based synthetic methods, including coprecipitation, the reverse micelle method, microwave plasma synthesis, sol-gel techniques, freeze drying, ultrasound irradiation, hydrothermal methods, laser pyrolysis techniques, and therm decomposition of organometallic and coordination compound [37-44]. The application of small iron oxide particles for *in vitro* diagnostics has been practiced for nearly 40 years [45]. In the last decade, investigations with several types of iron oxides have been increasingly carried out in the field of nanosized magnetic particles.

Since MNPs can be conducted by an external magnetic field their use in magnetic resonance imaging (MRI), tissue repair, hyperthermia, drug delivery, and in cell separation are very important [32-34, 46] (Fig. 1). All these applications require that the cell be efficiently captured by the magnetic nanoparticles either *in vitro* or *in vivo*. Unfortunately, the endocytotic internalization of nanoparticles into cells is severely limited by the short dwell time of these particles in the blood and non-specific targeting for achieving the sustained expression on levels required for these applications [47]. In an *in vivo* situation, macrophages of the reticuloendothelial system rapidly confront to and internalize MNPs, reducing their cytotoxic potential effect [46]. This can be an unlikable effect if their cytotoxic aspect is need, for example to kill tumorous cells. One strategy found to increase MNPs life time in human body is promoting chemical modifications on their surface [46].

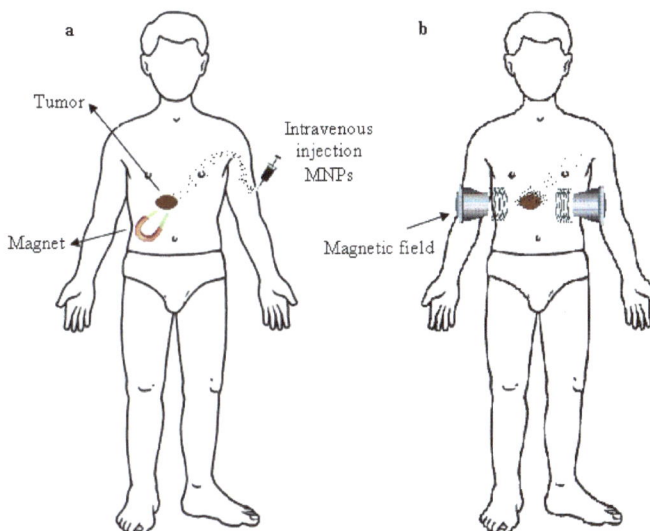

Figure 1: Some applications of MNPs: (a) MRI diagnosis, drug delivery, (b) hyperthermia.

Engineering surfaces of nanoparticles with polymers and proteins or coupling targeting ligands to these nanoparticles can confound the macrophages [46, 48], and further improve tissue selectivity [48].When the MNPs surfaces are modified by different agents such as drugs and bioactive agents, they are able to go into cells and tissue barriers and can release the drugs directly to the target organ promoting more efficient treatment or diagnose [34, 49]. Fig. **2** shows nanoparticles binded to the receptors on the cell surface and nanoparticles being internalized.

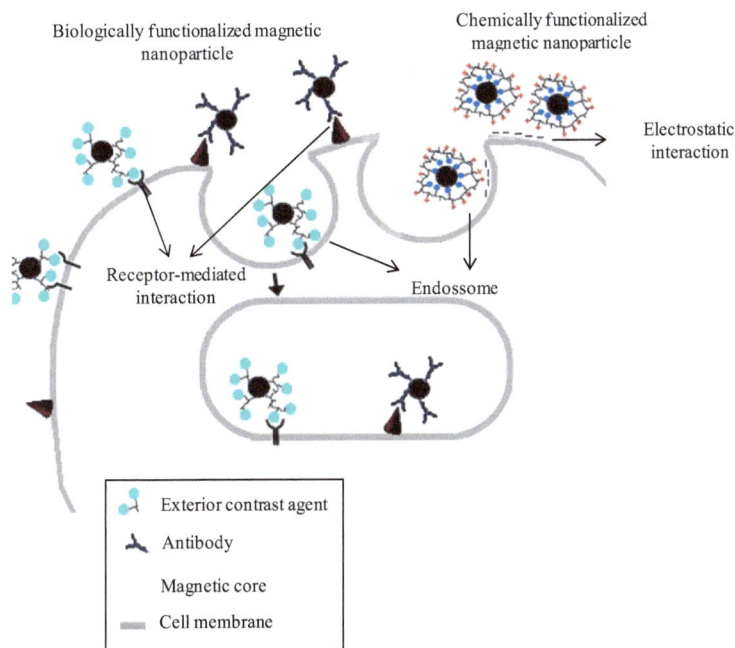

Figure 2: Nanoparticles binds to the receptors on the cell surface and nanoparticles being itself internalised.

Advances on nanomedicine using MNPs certainly demand more integrated studies on knowledge of surface property and its interface with the target organ. [6, 50-52]. The toxic effects of these functionalized nanomaterials need to be searched since the same microsized material effects are very unusual. The higher toxic potential of nanosized materials comparative to their bulk and molecular counterparts is due to their larger reactive surface area, ability to cross-cell and tissue barriers, and resistance to biodegradation [53], being all these aspects paradoxically usefull for their applications. Nanotoxicity is correlated to the the apoptosis and genotoxicity induced by oxidative stress and inflammatory steps [6, 51-54]. A number of reviews are focused on designs, devices, chemical and physiological properties and biomedical purposes of MNP [33, 34, 46, 55, 56]. The present review will deals with current studies and issues of MNP toxicity.

SURFACE CHEMISTRY AND BIOCOMPATIBILITY

As prepared, MNPs have large hydrophobic surface area which has raise susceptibility to agglomerate [55]. For the major variety of applications, the MNP surfaces need to be coated in order to be dispersed homogeneously and keep on these nanosized forms. Two mechanisms: electrostatic interation and steric interation are used to improve MNP stability. Illustration of these stabilization mechanisms are given in Fig. **3**. Examples of coating materials currently used to modify MNP surface include biodegradable polymers, inorganic metals, and inorganic oxides. Some examples of the using of these coating materials are summarized in Table **1**.

Table 1: Some polymers/molecules which can be used for nanoparticle coating to stabilize the ferrofluids and also for other biological applications.

Polymers / molecules	Refs.	Structural formulas
Dextran	[64, 65]	
Polyvinyl alcohol (PVA)	[66]	
Chitosan	[67]	
Polyethylene glycol (PEG)	[68-70]	
Polyvinylpyrrolidone (PVP)	[71]	

Fatty acids	[72]	With carboxyl groups (-COOH) binder to a big alkylic saturated or insatured chain
Polyacrylic acid	[73]	
Polypeptides	[74, 75]	Polypeptide corresponds to a linear polymer with more than 10 amino acids
Phosphorylcholine	[76]	
Poly (N-isopropylacrylamide) (PolyNIPAAM)	[77]	

BIODEGRADABLE POLYMERS

Polymer coated iron-oxide particles (SPIO and USPIO) minimize aggregation problems and this has allowed development of MNPs for various specific fields of clinical MR imaging [57-61]. In fact, all SPIO or USPIO MR contrast agents already approved for clinical usage nowadays, as well as most of the currently developing contrast agents, were stabilized by dextran or its derivatives. It was reported that the reason for the stability of dextran-coated iron oxide particles was that a COO$^-$ terminal of dextran bound to a Fe atom on the core surface [62]. Electrostatic repulsion between particles with the same electric charge prevents the aggregation of particles (Fig. **3**) [63].

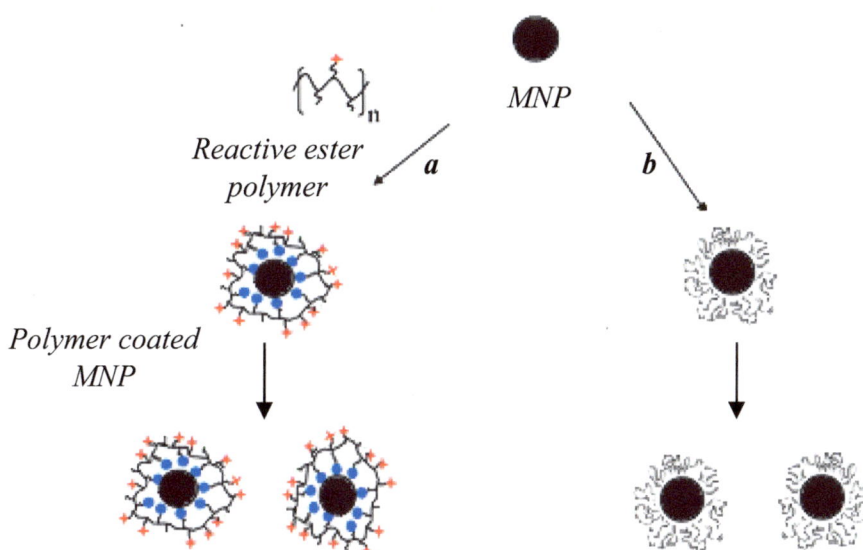

Figure 3: Illustration of suspension MNPs stabilization mechanisms: (a) Electrostatic interation, (b) Steric interation.

The polymer coating significantly increases their overall size and therefore may limit their tissue penetration ability, and metabolic clearance [78]. In general, the biodistribution of these polymer-based nanoparticles also is influenced by their size and surface chemistry [79, 80]. Thus, the Feridex™, a contrast-enhanced MRI of hepatic tumors, approved by US Food and Drug Administration (FDA) [81], with particles sized between 80 and 120 nm, is mainly trapped in the liver, while smaller sizes (about 30 nm) commonly are useful for imaging the lymph node systems (*e.g.,* Combidex™, Advanced Magnetics, Cambridge, MA; Sinerem™, Laboratoire Guerbet, F) [78]. In the surface chemistry aspect, since both the cell membrane and the iron oxide nanoparticles have a net negative charge, cells do not normally take up the nanoparticles. The positively charged dextran, used to coating the MPNs in Feridex™, however, forms a complex with the iron oxide nanoparticles, changing the electrostatic properties of the surfaces of the particles and causing cell uptake [81]. The iron oxide nanoparticles are then spontaneously internalized by phagocytes (Kupffer cells). On post-contrast MRI scans, the phagocyte-rich liver turn dark while the tumor, lacking macrophages, remains isointense. For magnetic prelabeling of nonphagocytic cells and cellular MRI, Feridex™ can be combined with other commercially available transfection agents such as a poly-L-lisina (PLL) [82].

Other factor to be considered in coating nanoparticles is the kind of desirable surface to produce. Hydrophobic surface may enhance the uptake of the nanoparticles by cells, while hydrophilic coating and surface charges may influence their retention period in the circulation and also the chance to penetrate into interstitial cell spaces [58-61, 79, 80, 83, 84]. Although these polymer coatings are generally considered to be biocompatible, adverse side-reactions have also been reported [85-89]. Moreover, many polymer-coated MNPs are unstable at high temperatures and are not suitable to protect reactive MNPs due to their poor air stability and high susceptibility to leaching under acidic conditions [55]. Poor functionalization and conjugation capacity of polymer-coated MNPs [34] limit their use for more complex applications. However, amination or carboxylation of dextran, resulting in cross-linked iron oxide MNPs, improves their conjugation capacity. For example, carboxymethyl dextran (CMD) is a polysaccharide with carboxyl groups that can be used as a spacer to covalently link bioactive compounds, and as a low-fouling surface-coating layer to limit non specific protein desorption [90]. These coatings could lead to the development of well-defined surface modifications that allow a precise control of interactions at the tissue-biomaterial interface, and ultimately improve the performance of long-term biomaterial implants [91]. CMD not only can stabilize the nanoparticle colloid by its carboxymethyl groups, but it can also provide suitable anchor groups for the covalent fixation of biomolecules. CMD-coated magnetic nanoparticles (CMD-MNPs) have been used to immobilize oligonucleotides for potential applications in tumor diagnosis and therapy [92].

Polyethylene glycol (PEG) is a polymer that is widely used for its hydrophilicity and low antigenicity. In addition to steric stabilization of MNPs, PEG prevents their plasma opsonization and uptake by macrophages, increasing MNP circulation *in vivo* [70]. PEG-coated MNPs are efficiently internalized by cells *via* fluid-phase endocytosis and through amphiphilic affinity to lipid bilayers on plasma membranes [70]. These qualities, however, increase their potential to overload cells with iron and become toxic [33]. Polyvinyl alcohol (PVA) has excellent film-forming, emulsifying and adhesive properties [33]. Hydroxyl groups from PVA are able to quell metallic ions and form a stable tridimensional structure [93]. While PVA use for intravenous delivery is limited due to poor persistence and agglomeration parameters in a ferrofluid [33], when cross-linked to form a magnetic gel, PVA can be used as a vitreous eye substitute [94]. Potentially, this property can be used in targeted drug delivery, tissue engineering and biosensor technology [33]. Chitosan, poly(1→4)-2-amino-2-deoxy-d-glucan, is a polyaminosaccharide with many significant biological (biodegradable, biocompatible and bioactive) and chemical properties (polycationic, hydrogel, reactive groups such as OH and NH_2). Thus, chitosan and its derivatives have been widely used in the fields of medicine, pharmacy and biotechnology [95]. It is suitable for affinity purification of proteins and magnetic bioseparation [33].

INORGANIC METALS

Precious metals, such as gold, have been used to protect iron oxide cores against oxidation, as they form highly stable particles of low reactivity [96, 97]. Nanogold is known for its superior optical properties, biocompatibility and outstanding capacity for functionalization [96, 97]. These properties make gold a particularly attractive surface chemistry for MNPs. One limitation is that gold coating can weaken magnetic properties of composites based on iron oxide MNPs [56] and is considered difficult to achieve due to the dissimilar nature of the two surfaces [55, 97]. When achieved, gold-coated MNPs are stable under neutral and acidic conditions [98, 99]. Engineering a continuous gold shell over an MNP core is a valuable challenge, as it can provide an extremely effective barrier against oxidizing agents [33]. In addition, the nanosized gold has shown a change in the localized surface plasmon band upon molecular adsorption, which might contribute to the development of novel biosensors based on this colorimetric change. These gold shells were fabricated either through chemical reduction of a gold precursor or by γ-ray irradiation [99, 100]. However, applications of the gold-coated magnetic nanomaterials to biological fields are relatively rare in spite of a wealth of reports on their manufacture. This might be due to difficulties to keep biomolecules active as well as to the needs of a specific chemical modification to prevent the nanomaterials from aggregating.

INORGANIC OXIDES

An inert silica shell of different thicknesses is used to tune the MNP core to different extents of dipolar coupling (*i.e.,* direct magnetic dipole-dipole interaction) and cooperative magnetic switching [55]. MNPs can be directly coated with amorphous silica produced by the hydrolysis of a sol-gel precursor [101, 102]. Because the iron oxide surface has a strong affinity to silica, no primer is required to promote silica deposition and adhesion [55]. The *in vitro* preliminary evaluation showed that silica-coated MNPs did not produce any severe alteration in biocompatibility [29].

Based on the negative charge of the silica shell, silica-coated MNPs are redispersible and stable in aqueous conditions [55]. They provide good control of interparticle interactions both in solutions and within structures through variations of the shell thickness [55]. Depending on the protocol for their synthesis, silica-coated iron oxide MNPs range in size from 1-2 nm [103] to 150 nm [104], but the development of monodispersed solutions often proves to be difficult. Silica-coated MNPs have longer circulation times and their hydrophilic negatively charged surface provides ideal anchorage for covalent binding to ligands presenting an excellent platform for drug delivery [33].

FUNCTIONALIZED MNPS

Core-shell-type composites consisting of magnetic nanoparticles decorated with biological substances are interesting for various biomedical applications. When designing such composites, the main challenge has mostly been to immobilize a given bioactive substance on the surface with a sufficiently strong bond to ensure the composite's stability prior to and during the biomedical application. For certain applications, however, it would be desirable for the bonding of the biosubstance to the magnetic core to be reversible. Both coated and uncoated MNPs have been functionalized with peptides, nucleic acids, small molecules and sometimes antibodies. Many examples of these applications are given in the scientific literature, as following:

Separation of Biological Actives Molecules

The protein separation with organosilane, such as carboxyl, aldehyde, amine, and thiol groups, and also assembled silica coated magnetic nanoparticles was achieved for model proteins such as bovine serum albumin (BSA) and lysozyme (LSZ) at different pH conditions. A work on using these amino functionalized silica coated magnetic nanoparticles for protein purification indicates that they have many advantages such as easy preparation, low cost, easy handling and rapid purification. These particles have extensive potential for serving as a very useful tool for facilitating biotechnology applications [105].

MNPs also have to be coated with amorphous silica shells for enhanced surface reactivity and RNA and DNA purification [106]. DNA is a polyanionic molecule due to the presence of phosphate groups on the nucleic acid backbone and is conveniently captured on a polymeric resin or other metal/inorganic supports with positively charged functional groups. However, current DNA purification methods suffer from several drawbacks that make them unsuitable for the manufacture of pharmaceutical grade. They often involve the use of solvents, toxic chemicals such as cesium chloride, ethidium bromide, phenol, and chloroform, or animal-derived enzymes such as ribonuclease A and lysozyme that are either not approved or not recommended by regulatory agencies. Finally, many techniques were designed to produce small quantities of DNA for laboratory use and are not suitable for the production of therapeutic materials at larger scale [106]. Surface modifications of magnetic particles with suitable intermediates are commonly used to extract the desired target. The driving forces for adsorption processes are hydrophobic, electrostatic, and ligand binding interactions [107, 108]. Desorptions of the biomolecules from the magnetic particles could be achieved by using high concentration salts, changing pH, and temperature, for undergoing conformational changes.

Functionalized Nanoparticles as Biosensors

Due to their biocompatible catalytic activity and low toxicity, it would be promising to utilize nano-Fe_3O_4 for biosensor applications [109]. Furthermore, it was reported that the immobilization of enzymes on magnetic nanoparticles could potentially result in unique properties of bioactive particles, such as increased enzyme activity due to the increased surface area of nanoparticles, and good dispersion in the analyte solution leading to rapid contact between the enzyme and its substrate and reduction of mass-transfer limitations [110]. In the study Lu *et al.* [111], magnetic nano-Fe_3O_4 were prepared by co-precipitation method [109] and a disposable glucose biosensor with high sensitivity was prepared by dropcoating of mixture of ferricyanide (Ferri)-Nano-Fe_3O_4 onto the surface of screen-printed carbon electrodes, and glucose oxidase (GOD) was then layered on. The glucose biosensors exhibit a relatively fast response (<15 s) and high sensitivity (ca. 1.74 μAmM^{-1}). The glucose biosensor has been widely used as a clinical indicator of diabetes and in the food industry for quality control [112, 113].

MNP ABSORPTION, METABOLISM, CLEARANCE AND TOXICITY

Bioabsorption of Suspended Particles in Air

The way airborne solid particles behave and their fate in the natural medium are conditioned by several intrinsic and environmental factors: initial particle dimension, chemical characteristics, time in suspension, inter-particles interactions and trajectory length in air. The main physical processes so involved are: particle diffusion, agglomeration, wet- and dry-deposition, and gravitational settling. All those processes, which are well established for ultrafine particles, may be also applicable to nanomaterials [114]; in few cases, nanomaterials may behave quite differently from bigger particles. For instance, coated nanoparticles tend to not exhibit agglomeration, but, for their sizes, are more likely inhaled, ingested, or absorbed. In addition, they may reach deep into the lung and are much more movable inside the body [54], and are not very efficiently phagocytized by the alveolar macrophages, which is the main mechanism for particle clearance in the alveolar region.

BIODISTRIBUTION

Nanotechnology has been increasingly employed in drug delivery. By nanosizing a formulation, we can increase the drug dissolution rate, leading to enhanced drug absorption and bioavailability. Using the nanoparticles to deliver drugs, improved tissue selectivity can be achieved due to the selective uptake of nanoparticles in certain tissues. The pharmacokinetics and tissue distribution of the nanoparticles largely define their therapeutic effect and toxicity. The physical-chemical aspects, as well as the route of inoculation can influence the pattern of distribution of MNP in the body [46]. Physiologically, about 80-90% of MNP injected intravenously are captured on the liver, spleen and 5-6% 1-2% are located in the bone marrow. On the other hand, the MNP inhaled particles accumulate in the brain, liver, spleen or lungs [115]. The particles size also influences the distribution *in vivo*, for example, particle size greater than 200 nm is sequestered in the spleen by phagocytosis. Particles smaller than 10 nm are rapidly removed in the kidneys [32]. Besides the mentioned organs, MNP particles are also found in hair follicles, cornea and epidermis [116]. This distribution described can be influenced by chemical treatment of MNP used that promotes activation and ingestion of particles by phagocytic cells such as macrophages, dendritic cells and endothelial cells located in these regions of the body [53, 54, 117-120]. At this moment, we do not known well, what the physiological and biochemical mechanisms involved in the degradation of MNP. So far, we know that after phagocytosis mediated by interactions with the mannose, components of the complement system, Fcγ receptor and scanvenger receptors and endocytosis (clatrin-and calceolin-mediated, fluid-phase) [121, 122], lysosome enzymes promote degradation in iron ions [33, 123]. In addition, particles of NPM can interact with plasma proteins. These particles are then detected as well opsonized and phagocytized by the reticuloendothelial system (RES) [122, 124, 125]. Researchs are being conducted to change the size and chemical surface composition of these particles in order to decrease the interaction with plasma proteins and thereby prevent phagocytosis. This measure, in a way, should increase the permanence time of particles in the circulation [121]. To do this, it has been used with great success, the coating of MNP with amphiphilic polymers surfactants based on polyethylene glycol (PEG). This treatment significantly reduced the particle interactions with plasma proteins and prevents phagocytosis by macrophages [46, 126-128].

MNPS METABOLISM INVOLVING MACROPHAGES

Macrophages are important cells in the body's defense system against diseases, and introduction of nanoparticles into the body may affect macrophage defensive function. Understanding the molecular interactions at the macrophage-nanoparticle interface has led to new approaches for selective MR imaging of cardiovascular diseases, multiple sclerosis, stroke and other diseases [129-133]. It is known that nanoparticles promote activation as well as phagocytotic, cytoskeletal and cytokine releasing functions of macrophages [122, 134-136]. Besides that, dextran-coated MNPs induce differentiation of monocytes into macrophages. While proinflammatory signaling may manifest macrophage activation, it may mediate MNP-induced macrophage cytotoxicity. Given that particle clearance by macrophages is size-dependent, ultra small nanoparticles can escape phagocytosis and lavage within the macrophages [137], promoting activation of OS and ROS (reactive oxygen species)-mediated redox-sensitive transcription *via* NF-κBand AP-1 [138], proinflammatory TNF-α /p38 signaling and apoptosis [136]. In fact, these properties have translated into therapeutic «macrophage-suicide» approaches of nanocarrier systems for macrophage-driven diseases of bacterial infection, atherosclerosis, rheumatoid arthritis and neuroinflammation [118, 124]. To understand the multi-factorial relationship of MNPs with macrophages, engineering strategies to increase the ability of MNPs to evade macrophages have to be accompanied by thorough evaluation of their toxic potential associated with relevant routes of delivery and identification of target cells of their internalization.

OXIDATIVE STRESS (OS): A PARADIGM FOR NANOTOXICITY

Nanoparticles could induce intracellular OS by disturbing the balance between the oxidant and antioxidant processes [6, 139]. Thus, MNPs are believed to induce redox cycling and catalytic chemistry *via* the Fenton reaction $[H_2O_2 + Fe^{2+} \rightarrow Fe^{3+} + HO^- + HO^*]$ [6], the most prevalent source of ROS in biological systems. In fact, uncoated magnetite nanoparticles are significantly cytotoxic [70], but the mechanism of cytotoxicity of the fully oxidized (Fe_2O_3) maghemite MNPs is not directly explained by the Fenton chemistry. Oxidative Stress (OS): A Paradigm for Nanotoxicity OS and reactive oxygen species (ROS) formation were shown to be some of the key mechanisms in cellular defense after particle uptake. Furthermore, ROS generation leading to OS is associated with the nanotoxicity of nonmetal nanoparticles as well [6, 53, 54, 117, 140]. While the mechanisms by which nanoparticles generate ROS are still unknown, it is hypothesized that disruption of the well-structured electronic configuration of the nanosized material surface creates reactive electron donor or acceptor sites, leading to the formation of superoxide radicals [6]. This process results in the production of proinflammatory cytokine, chemokine and matrix metalloproteinase (MMP) release (tier II OS), *via* MAPK signaling pathways, which leads to apoptosis (tier III OS) [6, 121, 137, 141]. Tier II OS can manifest *in vivo* through immune cell infiltration, mediated by the actions of proinflammatory cytokines and MMPs [142]. It was demonstrated that ROS controls MMP activity by two distinct mechanisms: MAPK-induced over expression of MMP gene and directly through structural oxidative modification of thiol residues on inactive pro-MMPs, resulting in release of the zinc-binding domain and MMP activation [143].

METALS AND MNPS IN NEURODEGENERATION

There is evidence that the induction of ROS activity of MNP in the nervous system can lead to increased permeability of blood-brain barrier, and neuronal damage in consequence of the activation of macrophages by nanoparticles [124, 144-146]. In the context of MNP induced macrophage recruitment into the nerve, it is worth noting that cytokines and MMPs are important modulators of macrophage recruitment into the nerve [147, 148]. Together, these studies support a model in which MNPs induce ROS to promote MMP mediated degradation of the blood-brain and blood-nerve barriers to promote recruitment of macrophages into the nervous system. Although iron oxides are in general considered to be safe, imbalance in its homeostasis has known toxic implications to many organ systems [149]. As the abundant, redox-active metal, iron facilitates the generation of free radicals in the brain [150, 151] and excess iron is associated with multiple neurodegenerative disorders, including multiple sclerosis, Alzheimer's and Parkinson's diseases [152]. Nanoscale iron oxide deposits have been identified in neurological tissues using SQUID magnetometry in the brains of Alzheimer's patients [153] and basal ganglia of neuroferritinopathy patients [154]. These iron nanoparticles are believed to be biogenic in nature and linked to ferritin, a large (12 nm) natural ferric oxide phosphate storage protein [54, 153, 155]. However, evidence of brain accumulation of MNPs after passive exposure through inhalation [115] calls for confirmation of the source of MNP deposits in the brains of neurodegeneration patients. It is particularly important due to the ability of MNPs to penetrate through skin [116] and cross blood-brain barrier after exposure through inhalation [115] and by intraperitoneal delivery [156]. Research on the neurotoxicity of MNPS and the identification of formulations for use without promoting toxicity are needed before its therapeutic use. As an example quoted of iron chelation therapy used to neurodegenerative diseases [157] and drug conjugates (eg, glial-derived neurotrophic factor) as a model of therapy for the treatment of drug addicts [158].

CURRENT STUDIES ON MNP TOXICITY

Current approaches strongly suggest that consequences of nanotechnology are best addressed within the existing system applications such as biology, chemistry, or electronics. Nanoparticles can be produced from nearly any chemical. However, most nanoparticles that are currently being used are made from transition metals, silicon, carbon (single-walled carbon nanotubes), and metal oxides. Meanwhile, potential public and occupational exposures to manufactured nanoparticles should increase dramatically, because nanomaterials are supposed to improve the quality and performance of many consumer products as well as medical therapies. Therefore, it is time that information regarding the risk assessment of manufactured nanoparticles is collected. Critical questions regarding the potential human health and environmental impact of manufactured nanoparticles or nanomaterials have been raised only recently. The use of particle reduction in size from micro to nanoscale not only provides benefits to diverse scientific fields but also poses potential risks to humans and the environment. For the successful application of nanomaterials in bioscience, it is essential to understand the biological fate and potential toxicity of nanoparticles. Such a rapid development in nanotechnology will result in several changes in areas such as nanoscale visualization, insight into living systems, revolutionary biotechnology, synthesis of new drugs as their targeted delivery, and regenerative medicine and offer many other benefits. Further, nanomaterials can be modified for better efficiency to facilitate their applications in different fields such as bioscience and medicine. For example, magnetic nanoparticles (MNPs) have been studied with intent to apply them in bioscience, because they offer benefits such as separation and gathering of materials of interest by using a magnetic force. However, the lack of information regarding the toxicity of manufactured nanoparticles poses serious problems. Physical and chemical characteristics of nanoparticles (size, shape, surface chemistry, solubility) each influence their toxic activity [159]. Particokinetics and dosimetry approaches to nanotoxicity assessment encompass measures for dose of exposure (mass administered, media mass, surface area), delivered dose (number per cell or cm^3), and cellular dose (internalized mass) [160]. With many parameters to assess, the demand for toxicity evaluation exceeds current capabilities of the research field. Reducing the particle size of materials is an efficient and reliable tool for improving the bioavailability of a gene or drug delivery system. In fact, nanotechnology helps in overcoming the limitations of size and can change the outlook of the world regarding science. Further, nanomaterials can be modified for better efficiency to facilitate their applications in different fields such as bioscience and medicine. Thus, prioritization of the physicochemical parameters of clinical relevance to the target organ groups is imperative. *E.g.,* for skin, that includes nanoparticle dissolution, size and partition coefficient, whereas for brain, hydrophobicity and surface chemistry, relating to possibility to cross barriers, size and shape and chemical composition are more relevant [161]. It has been suggested that study designs center only on clinically relevant target cells and include in-depth mechanistic analyses of the hierarchical process of OS [6]. For tier I OS, ROS species generation and measures of anti-oxidative HO^{-1} have been proposed as useful measures of OS-specific nanotoxicity [136]. As mentioned already, activation of proinflammatory signaling and MMPs have been adopted by many studies and represent reliable measures of tier II OS. It has been shown, however, that measures of MMP-9 provide higher sensitivity when cytokine measures fail [124]. However, the lack of information regarding the toxicity of manufactured nanoparticles poses serious problems. Therefore, it is necessary that specialists and researchers in toxicology, chemistry, and other fields are aware of the importance of analyzing the positive aspects of nanomaterials while avoiding their potential toxic effects. Overall, the use of multiple assays is important to avoid false-positive or false-negative results [162]. Tier III OS can be identified by activation of pro-apoptotic pathways [6, 136]. Thus, whether or not activation of tier I OS would indicate ROS-induced changes of nanotoxicity, determination of cell apoptosis of tier III OS will help understand the extent of cytotoxicity. *In vivo*, the cell source for specific changes could be identified by dual-labeling of OS-specific antigens and cell specific markers using confocal microscopy. Caution in the interpretation of mechanistic studies is important and a consideration that low-level ROS signaling contributes to normal cellular redox signaling [163] and cytokine-protease activation may relate to the body's defense reaction in immobilizing immune cells. The biokinetics of nanoparticles through confirmed and potential routes of exposure involves multiple pathways of their translocation prior to clearance [54], particularly as nanoparticle binding to proteins generates complexes that are more mobile and pervasive through tissues normally inaccessible [6]. The time-course of nanotoxicity measures may help reconcile studies demonstrating safety of MNPs [156, 164, 165] with those of transient and acute *in vivo* toxicity [156, 166]. Some toxicity data may not be specific to nanoparticles per se, but relate to the increased substance quantities and accumulation, which could be adjusted with adoption of appropriate concentration standards [161]. Since most substances become toxic at high doses, it is important that nanotoxicity studies incorporate doses at the anticipated human exposures [117]. Toxicity could also be qualitatively different, basing on size, surface chemistry or specific interaction [161]. When nanotoxicity is observed, the role of each component of the composite structure should be evaluated as control factors, if possible (*e.g.,* its vehicle or coating material). For example, polyakylcyaniacrylate- coated MNPs (220 nm, with 10-20 nm cores) showed an LD50 of 245 mg/kg, however, a similar LD50 was observed in polyakylcyaniacrylate (not MNP-based) particles [167]. As mentioned, MNP surface chemistry that ensures high efficiency of intracellular MNP uptake or effective evasion of the RES is likely to increase their toxic potential [33]. Although the studies on MNP toxicity are limited in number, they point at the influence of MNP composition, size, dispersibility, surface chemistry and the regimens of their administration in the toxic outcomes as anticipated [6].

CONCLUSIONS

Nano-materials containing magnetic particles are particularly important in clinical diagnosis and therapeutical medicine, for their singular physical and chemical properties, mainly related to their size. Many industrial and biomedical applications result from characteristics related to extremely reduced sizes, which confer mobility, hyperthermy due to superparamagnetic behavior and attraction response to magnetic fields to these materials. They are also of relatively easy preparation and are of low cost, particularly those based on iron oxides. However, the area surface of these materials is relatively so large that they are very reactive to agglomeration, an unwished process that tends to occur very quickly. In this review, many topics on the dispersion of magnetic nanomaterials in air and in the human body, which may address many of their metabolic toxicity effect, and eventually limit their medical use, are discussed. To prevent these unwanted side effects, it is necessary to develop new or optmize current methods used to coat nano-surfaces and to help preserve the planned initial characteristics of such nanosized materials. The choice of the coating method may lead to materials with functionalized surfaces, able them to be bound to specific human organs or tissues. Nevertheless, as in each new research area, many studies must still be made so that gives safer nanoscale materials to be manipulated and implanted into human bodies.

ACKNOWLEDGEMENT

This work supported by FAPEMIG (Brazil).

REFERENCES

[1]   Andrade AL, Souza DM, Pereira MC, Fabris JD, Domingues RZ. Catalytic effect of magnetic nanoparticles over the H_2O_2 decomposition reaction. J. Nanosci Nanotech 2009; 9: 3695-9.

[2]   Anselmann R. Nanoparticles and nanolayers in commercial applications. J Nanoparticle Res 2001; 3: 329-36.

[3]   Doumanidis H. The nanomanufacturing programme at the National Science Foundation. Nanotechnology 2002; 13: 248-52.

[4]   Emerich DF, Thanos CG. Nanotechnology and medicine. Expert Opin Biol Ther 2003; 3: 655-63.

[5]   Lowe T. The revolution in nanometals. Adv Mater Process 2002; 160: 63-5.

[6]   Nel A, Xia T, Madler L, Li N. Toxic potential of materials at the nanolevel. Science 2006; 311: 622-7.

[7]   Shinkai M. Functional magnetic particles for medical application. J Biosci Bioeng 2002; 94: 606-13.

[8]   Gould P. Nanomaterials face control measures. Nano today 2006; 1: 34-9.

[9]   Karlsson HL, Cronholm P, Gustafsson J, Moller L. Copper oxide nanoparticles are highly toxic: A comparison between metal oxide nanoparticles and carbon nanotubes. Chem Res Toxicol 2008; 21: 1726-32.

[10]   Nurmi JT, Tratnyek PG, Sarathy V, *et al.* Characterization and properties of metallic iron nanoparticles: Spectroscopy, electrochemistry, and kinetics. Environ Sci Technol 2005; 39: 1221-30.

[11]   Zhang WX. Nanoscale iron particles for environmental remediation: An overview. J Nanoparticle Res 2003; 5: 323-32.

[12]   Soto K, Garza KM, Murr LE. Cytotoxic effects of aggregated nanomaterials. Acta Biomater 2007; 3: 351-8.

[13]   Warheit DB, Webb TR, Reed KL, Frerichs S, Sayes CM. Pulmonary toxicity study in rats with three forms of ultrafine-TiO_2 particles: Differential responses related to surface properties. Toxicology 2007; 230: 90-104.

[14]   Chen HW, Su SF, Chien CT, *et al.* Titanium dioxide nanoparticles induce emphysema-like lung injury in mice. FASEB J 2006; 20: 2393-5.

[15]   Soto KF, Carrasco A, Powell TG, Garza KM, Murr LE. Comparative *in vitro* cytotoxicity assessment of some manufactured nanoparticulate materials characterized by transmission electron microscopy. J Nanopart Res 2005; 7: 145-69.

[16]   Dreher KL. Health and environmental impact of nanotechnology: Toxicological assessment of manufactured nanoparticles. Toxicol Sci 2004; 77: 3-5.

[17]   Swiss Report Reinsurance Company. 2004. Nanotechnology: Small Matter, Many Unknowns. www.swissre.com. Accessed on 21 July 2010.

[18]   UK Royal Society. 2004. The Royal Society and the Royal Academy of Engineering. Nanoscience and Nanotechnologies: Opportunities and Uncertainties. http://www.nanotec.org.uk/finalreport.htm. Accessed on 21 July 2010.

[19]   European Commission. 2004. European Commission, Community Health and Consumer Protection. Nanotechnologies: A Preliminary Risk Analysis on the Basis of a Workshop Organized in Brussels on 1-2 March 2004 by the Health and Consumer Protection Directorate General of the European Commission. http://europa.eu.int/comm/health/ph_risk/events_risk_en.htm. Accessed on 21 July 2010.

[20]   European NanoSafe Report. 2004. Technical Analysis: Industrial Application of Nanomaterials Chances and Risks. www.nano.uts.edu.au/nanohouse/nanomaterials%20risks.pdf. Accessed on 21 July 2010.

[21]   Pulskamp K, Diabate S, Krug HF. Carbon nanotubes show no sign of acute toxicity but induce intracellular reactive oxygen species in dependence on contaminants. Toxicology Letters 2007; 168: 58-74.

[22]   Morgan K. Development of a preliminary framework for informing the risk analysis and risk management of nanoparticles. Risk Analysis 2005; 25:1621-35.

[23]   Holsapple MP, Farland WH, Landry TD, *et al.* Research strategies for safety evaluation of nanomaterials, Part II: Toxicological and safety evaluation of nanomaterials, current challenges and data needs. Toxicol Sci 2005; 88: 12-7.

[24]   Balshaw DM, Philbert M, Suk WA. Research strategies for safety evaluation of nanomaterials, Part III: Nanoscale technologies for assessing risk and improving public health. Toxicol Sci 2005; 88: 298-306.

[25]   Tsuji JS, Maynard AD, Howard PC, *et al.* Research strategies for safety evaluation of nanomaterials, Part IV: Risk assessment of nanoparticles. Toxicol Sci 2006; 89: 42-50.

[26]   Borm P, Klaessig FC, Landry TD, *et al.* Research strategies for safety evaluation of nanomaterials, Part V: Role of dissolution in biological fate and effects of nanoscale particles. Toxicol Sciences 2006; 90: 23-32.

[27]   Powers KW, Brown SC, Krishna VB, Wasdo SC, Moudgil BM, Roberts SM. Research strategies for safety evaluation of nanomaterials, Part VI: Characterization of nanoscale particles for toxicological evaluation. Toxicol Sci 2006; 90: 296-303.

[28]   Thomas K, Sayre P. Research strategies for safety evaluation of nanomaterials, Part I: Evaluating the human health implications of exposure to nanoscale materials. Toxicol Sci 2005; 87: 316-21.

[29]   Souza DM, Andrade AL, Fabris JD, *et al.* Synthesis and *in vitro* evaluation of toxicity of silica-coated magnetite nanoparticles. J Non-Cryst Solids 2008; 354: 4894-7.

[30]   Xia T, Kovochich M, Liong M, *et al.* Comparison of the mechanism of toxicity of zinc oxide and cerium oxide nanoparticles based on dissolution and oxidative stress properties. ACS Nano 2008; 2: 2121-34.

[31]   Mahmoudi M, Simchi A, Milani AS, Stroeve P. Cell toxicity of superparamagnetic iron oxide nanoparticles. J Colloid Interface Sci 2009; 336: 510-8.

[32]   Gupta AK, Gupta M. Synthesis and surface engineering of iron oxide nanoparticles for biomedical applications. Biomaterials 2005; 26: 3995-4021.

[33]   Gupta AK, Naregalkar RR, Vaidya VD, Gupta M. Recent advances on surface engineering of magnetic iron oxide nanoparticles and their biomedical applications. Nanomed 2007; 2: 23-39.

[34]   McCarthy JR, Kelly KA, Sun EY, Weissleder R. Targeted delivery of multifunctional magnetic nanoparticles. Nanomed 2007; 2: 153-67.

[35]   Tartaj P, Morales MD, Veintemillhas-Verdaguer S, Gonzalez-Carreno T, Serna CJ. The preparation of magnetic nanoparticles for applications in biomedicine. J Phys D-Appl Phys 2003; 36: R182-97.

[36]   Ito A, Shinkai M, Honda H, Kobayashi T. Medical application of functionalized magnetic nanoparticles. J Biosci Bioeng 2005; 100: 1-11.

[37]   Andrade AL, Souza DM, Pereira MC, Fabris JD, Domingues RZ. pH effect on the synthesis of magnetite nanoparticles by the chemical reduction-precipitation method. Quim Nova 2010; 33: 524-7.

[38]   Sugimoto T, Matijevic E. Formation of uniform spherical magnetite particles by crystallization from ferrous hydroxide gels. J Colloid Interface Sci 1980; 74: 227-43.

[39]   Sun SH, Zeng H, Robinson DB, *et al.* Monodisperse MFe_2O_4 (M = Fe, Co, Mn) nanoparticles. J Am Chem Soc 2004; 126: 273-9.

[40]   Elkins KE, Vedantam TS, Liu JP, *et al.* Ultrafine FePt nanoparticles prepared by the chemical reduction method. Nano Lett 2003; 3: 1647-9.

[41]   Jolivet JP, Chaneac C, Tronc E. Iron oxide chemistry. From molecular clusters to extended solid networks. Chem Commun 2004; 5: 481-7.

[42]   Li F, Liu J, Evans DG, Duan X. Stoichiometric synthesis of pure MFe_2O_4 (M = Mg, Co, and Ni) spinel ferrites from tailored layered double hydroxide (hydrotalcite-like) precursors. Chem Mater 2004; 16: 1597-602.

[43]   Sahoo Y, Cheon M, Wang S, Luo H, Furlani EP, Prasad PN. Field-directed self-assembly of magnetic nanoparticles. J Phys Chem B 2004; 108: 3380-3.

[44]   Song Q, Zhang ZJ. Shape control and associated magnetic properties of spinel cobalt ferrite nanocrystals. J Am Chem Soc 2004; 126: 6164-8.

[45]   Gilchrist RK, Medal R, Shorey WD, Hanselman RC, Parrot JC, Taylor CB. Selective inductive heating of lymph nodes. Ann Surg 1957; 146: 596-606.

[46]   Duguet E, Vasseur S, Mornet S, Devoisselle JM. Magnetic nanoparticles and their applications in medicine. Nanomed 2006; 1: 157-68.

[47]   Stolnik S, Illum L, Davis SS. Long circulating microparticulate drug carriers. Adv Drug Del Rev 1995; 16: 195-214.

[48]   Moghimi SM, Hunter AC, Murray JC. Long-circulating and target-specific nanoparticles: Theory to practice. Pharm Rev 2001; 53: 283-318.

[49]   Ozdemir V, Williams-Jones B, Glatt SJ, Tsuang MT, Lohr JB, Reist C. Shifting emphasis from pharmacogenomics to theragnostics. Nat Biotechnol 2006; 24: 942-7.

[50]   Powell MC, Kanarek MS. Nanomaterial health effects - Part 2: Uncertainties and recommendations for the future. WMJ 2006; 105: 18-23.

[51]   Powell MC, Kanarek MS. Nanomaterial health effects - Part 1: Background and current knowledge. Wisc Med J 2006; 105: 16-20.

[52]   Moore MN. Do nanoparticles present ecotoxicological risks for the health of the aquatic environment? Environ Int 2006; 32: 967-76.

[53]   Donaldson K, Stone V, Tran CL, Kreyling W, Borm PJA. Nanotoxicology. Occup Environ Med 2004; 61: 727-8.

[54]   Oberdorster G, Oberdorster E, Oberdorster J. Nanotoxicology: An emerging discipline evolving from studies of ultrafine particles. Environ Health Perspect 2005; 113: 823-39.

[55]   Lu AH, Salabas EL, Schuth F. Magnetic nanoparticles: Synthesis, protection, functionalization, and application. Angew Chem, Int Ed Engl 2007; 46: 1222-44.

[56]   Huber DL. Synthesis, properties, and applications of iron nanoparticles. Small 2005; 1: 482-501.

[57]   Koenig SH, Brown RD, 3rd. Relaxometry of magnetic resonance imaging contrast agents. Magn Reson Annu, 1987; 263-86.

[58]   Lind K, Kresse M, Debus NP, Muller RH. A novel formulation for superparamagnetic iron oxide (SPIO) particles enhancing MR lymphography: comparison of physicochemical properties and the *in vivo* behaviour. J Drug Target 2002; 10: 221-30.

[59]   Bellin MF, Beigelman C, Precetti-Morel S. Iron oxide-enhanced MR lymphography: initial experience. Eur J Radiol 2000; 34: 257-64.

[60]   Bonnemain B. Superparamagnetic agents in magnetic resonance imaging: Physicochemical characteristics and clinical applications. A review. J Drug Target 1998; 6: 167-74.

[61]   Muller RN, Vallet P, Maton F, *et al.* Recent developments in design, characterization, and understanding of MRI and MRS contrast media. Invest Radiol 1990; 25: S34-6.

[62]   Kawaguchi T, Hanaichi T, Hasegawa M, Maruno S. Dextran-magnetite complex: conformation of dextran chains and stability of solution. J Mater Sci Mater Med 2001; 12: 121-7.

[63]   Kumar MNVR, Bakowsky U, Lehr CM. Preparation and characterization of cationic PLGA nanospheres as DNA carriers. Biomaterials 2004; 25: 1771-7.

[64]   Berry CC, Curtis ASG. Functionalisation of magnetic nanoparticles for applications in biomedicine. J Phys D: Appl Phys 2003; 36: R198-206.

[65]   Berry CC, Wells S, Charles S, Curtis ASG. Dextran and albumin derivatised iron oxide nanoparticles: in.uence on .broblasts *in vitro*. Biomaterials 2003; 24: 4551-7.

[66]   Shan GB, Xing JM, Luo MF, Liu HZ, Chen JY. Immobilization of Pseudomonas delafieldii with magneticpolyvinyl alcohol beads and its application in biodesulfurization. Biotechnol Lett 2003; 25: 1977-81.

[67]   Khor E, Lim LY. Implantable applications of chitin and chitosan. Biomaterials 2003; 24: 2339-49.

[68]   Gupta AK, Wells S. Surface modified superparamagnetic nanoparticles for drug delivery: preparation, characterization and cytotoxicity studies. IEEE Trans Nanobiosci 2004; 3: 66-73.

[69]   Zhang Y, Kohler N, Zhang MQ. Surface modification of superparamagnetic magnetite nanoparticles and their intracellular uptake. Biomaterials 2002; 23: 1553-61.

[70]   Gupta AK, Curtis ASG. Surface modified superparamagnetic nanoparticles for drug delivery: interaction studies with human fibroblasts in culture. J Mater Sci Mater Med 2004; 15: 493-6.

[71]   D'Souza AJ, Schowen RL, Topp EM. Polyvinylpyrrolidone-drug conjugate: synthesis and release mechanism. J Cont Rel 2004; 94: 91-100.

[72]   Sahoo Y, Pizem H, Fried T, *et al.* Alkyl phosphonate/phosphate coating on magnetite nanoparticles: a comparison with fatty acids. Langmuir 2001; 17: 7907-11.

[73]   Burugapalli K, Koul V, Dinda AK. Effect of composition of interpenetrating polymer network hydrogels based on poly (acrylic acid) and gelatin on tissue response: a quantitative *in vivo* study. J Biomed Mater Res 2004; 68A: 210-8.

[74]   Lewin M, Carlesso N, Tung C-H, *et al.* Tat peptide-derivatized magneticnanopartic les allow *in vivo* tracking and recovery of progenitor cells. Nat Biotechnol 2000; 18: 410-4.

[75]   Bhadriraju K, Hansen LK. Hepatocyte adhesion, growth and differentiated function on RGD-containing proteins. Biomaterials 2000; 21: 267-72.

[76]   Denizot B, Tanguy G, Hindre F, Rump E, Jeune J, Jallet P. Phosphorylcholine coating of iron oxide nanoparticles. J Coll Interf Sci 1999; 209: 66-71.

[77]   Chen G, Hoffman AS. Preparation and properties of thermoreversible, phase separating enzyme-oligo(N-isopropylacrylamide) conjugates. Bioconjug Chem 1993; 4: 509-14.

[78]   Cheng FY, Su CH, Yang YS, *et al.* Characterization of aqueous dispersions of Fe_3O_4 nanoparticles and their biomedical applications. Biomaterials 2005; 26: 729-38.

[79]   Weissleder R, Elizondo G, Wittenberg J, Rabito CA, Bengele HH, Josephson L. Ultrasmall superparamagnetic iron oxide: characterization of a new class of contrast agents for MR imaging. Radiology 1990; 175: 489-93.

[80]   Thode K, Luck M, Schroder W, *et al.* The influence of the sample preparation on plasma protein adsorption patterns on polysaccharide-stabilized iron oxide particles and N-terminal microsequencing of unknown proteins. J Drug Target 1997; 5: 35-43.

[81]   Unger EC. Science to practice: How can superparamagnetic iron oxides be used to monitor disease and treatment? Radiology 2003; 229: 615-6.

[82]   Frank JA, Miller BR, Arbab AS, *et al.* Clinically applicable labeling of mammalian and stem cells by combining superparamagnetic iron oxides and transfection agents. Radiology 2003; 228: 480-7.

[83]   Reimer P, Kwong KK, Weisskoff R, Cohen MS, Brady TJ, Weissleder R. Dynamic signal intensity changes in liver with superparamagnetic MR contrast agents. J Magn Reson Imaging 1992; 2:177-81.

[84]   Chouly C, Pouliquen D, Lucet I, Jeune JJ, Jallet P. Development of superparamagnetic nanoparticles for MRI: effect of particle size, charge and surface nature on biodistribution. J Microencapsul 1996; 13: 245-55.

[85]   Reimer P, Balzer T. Ferucarbotran (Resovist): a new clinically approved RES-specific contrast agent for contrast-enhanced MRI of the liver: properties, clinical development, and applications. Eur Radiol 2003; 13: 1266-76.

[86]   Atkinson TP, Smith TF, Hunter RL. *In vitro* release of histamine from murine mast-cells by block co-polymers composed of polyoxyethylene and polyoxypropylene. J Immunol 1988; 141: 1302-6.

[87]   Ennis M, Lorenz W, Gerland W. Modulation of histamine-release from rat peritoneal mast-cells by non-cytotoxic concentrations of the detergents Cremophor El (oxethylated castor-oil) and Triton X100. A possible explanation for unexpected adverse drug reactions. Agents Actions 1986; 18: 235-8.

[88]   McLachlan SJ, Morris MR, Lucas MA, *et al.* Phase I clinical evaluation of a new iron oxide MR contrast agent. J Magn Reson Imaging 1994; 4: 301-7.

[89]   Nolte H, Carstensen H, Hertz H. VM-26 (teniposide)-induced hypersensitivity and de-granulation of basophils in children. Am J Pediatr Hematol Oncol 1988; 10: 308-12.

[90]   Huang BR, Chen PY, Huang CY, *et al.* Bioavailability of magnetic nanoparticles to the brain. J Magn Magn Mater 2009; 321: 1604-9.

[91]   Mclean KM, Johnson G, Chatelier RC, Beumer GJ, Steele JG, Griesser HJ. Method of immobilization of carboxymethyl-dextran affects resistance to tissue and cell colonization. Colloids Surf B Biointerfaces 2000; 18: 221-34.

[92]   Wagner K, Kautz A, Roder M, *et al.* Synthesis of oligonucleotide-functionalized magnetic nanoparticles and study on their *in vitro* cell uptake. Appl Organomet Chem 2004; 18: 514-9.

[93]   Lin H, Watanabe Y, Kimura M, Hanabusa K, Shirai H. Preparation of magnetic poly(vinyl alcohol) (PVA) materials by *in situ* synthesis of magnetite in a PVA matrix. J Appl Polym Sci 2003; 87: 1239-47.

[94]   Maruoka S, Matsuura T, Kawasaki K, *et al.* Biocompatibility of polyvinylalcohol gel as a vitreous substitute. Curr Eye Res 2006; 31: 599-606.

[95]   Ngah WSW, Ghani SA, Kamari A. Adsorption behaviour of Fe(II) and Fe(III) ions in aqueous solution on chitosan and cross-linked chitosan beads. Bioresour Technol 2005; 96: 443-50.

[96]   Daniel MC, Astruc D. Gold nanoparticles: assembly, supramolecular chemistry, quantum-size-related properties, and applications toward biology, catalysis, and nanotechnology. Chem Rev 2004; 104: 293-346.

[97]   Eustis S, El-Sayed MA. Why gold nanoparticles are more precious than pretty gold: Noble metal surface plasmon resonance and its enhancement of the radiative and nonradiative properties of nanocrystals of different shapes. Chem Soc Rev 2006; 35: 209-17.

[98]   Chen M, Yamamuro S, Farrell D, Majetich SA. Gold-coated iron nanoparticles for biomedical applications. J Appl Physi 2003; 93: 7551-3.

[99]   Lin J, Zhou WL, Kumbhar A, *et al.* Gold-coated iron (Fe@Au) nanoparticles: Synthesis, characterization, and magnetic field-induced self assembly. J Solid State Chem 2001; 159: 26-31.

[100] Seino S, Kinoshita T, Otome Y, *et al.* Gamma-ray synthesis of magnetic nanocarrier composed of gold and magnetic iron oxide. J Magn Magn Mater 2005; 293: 144-50.

[101] Stober W, Fink A, Bohn E. Controlled growth of monodisperse silica spheres in micron size range. J Colloid Interface Sci 1968; 26: 62-9.

[102] Andrade AL, Souza DM, Pereira MC, Fabris JD, Domingues RZ. Synthesis and characterization of magnetic nanoparticles coated with silica through a sol-gel approach. Cerâmica 2009; 55: 420-4.

[103] Santra S, Tapec R, Theodoropoulou N, Dobson J, Hebard A, Tan WH. Synthesis and characterization of silica-coated iron oxide nanoparticles in microemulsion: The effect of non-ionic surfactants. Langmuir 2001; 17: 2900-6.

[104] Tartaj P, Gonzalez-Carreno T, Serna CJ. Single-step nanoengineering of silica coated maghemite hollow spheres with tunable magnetic properties. Adv Mater 2001; 13: 1620-4.

[105] Chang JH, Kang KH, Choi J, Jeong YK. High efficiency protein separation with organosilane assembled silica coated magnetic nanoparticles. Superlattices and Microstructures 2008; 44: 442-8.

[106] Park ME, Chang JH. High throughput human DNA purification with aminosilanes tailored silica-coated magnetic nanoparticles. Materials Science and Engineering C 2007; 27: 1232-5.

[107] Massart R, Cabuil V. Effect of some parameters on the formation of colloidal magnetite in alkaline-medium - yield and particle-size control. Journal de Chimie Physique et de Physico-Chimie Biologique 1987; 84: 967-73.

[108] Donselaar LN, Philipse AP, Suurmond J. Concentration-dependent sedimentation of dilute magnetic fluids and magnetic silica dispersions. Langmuir 1997; 13: 6018-25.

[109] Chen DH, Liao MH. Preparation and characterization of YADH-bound magnetic nanoparticles. J Mol Catal B Enzym 2002; 16: 283-91.

[110] Rossi LM, Quach AD, Rosenzweig Z. Glucose oxidase-magnetite nanoparticle bioconjugate for glucose sensing. Anal Bioanal Chem 2004; 380: 606-13.

[111] Lu BW, Chen WC. A disposable glucose biosensor based on drop-coating of screen-printed carbon electrodes with magnetic nanoparticles. J Magn Magn Mater 2006; 304: e400-2.

[112] Tian FM, Zhu GY. Bienzymatic amperometric biosensor for glucose based on polypyrrole/ceramic carbon as electrode material. Analytica Chimica Acta 2002; 451: 251-8.

[113] Xu JJ, Yu ZH, Chen HY. Glucose biosensors prepared by electropolymerization of p-chlorophenylamine with and without Nafion. Analytica Chimica Acta 2002; 463: 239-47.

[114] Wiesner MR, Lowry GV, Alvarez P, Dionysiou D, Biswas P. Assessing the risks of manufactured nanomaterials. Environ Sci Technol 2006; 40: 4336-45.

[115] Kwon JT, Hwang SK, Jin H, *et al.* Body distribution of inhaled fluorescent magnetic nanoparticles in the mice. J Occup Health 2008; 50: 1-6.

[116] Baroli B, Ennas MG, Loffredo F, Isola M, Pinna R, Lopez-Quintela MA. Penetration of metallic nanoparticles in human full-thickness skin. J Invest Dermatol 2007; 127: 1701-12.

[117] Oberdorster G, Stone V, Donaldson K. Toxicology of nanoparticles: a historical perspective. Nanotoxicology 2007; 1: 2-25.

[118] Vega-Villa KR, Takemoto JK, Yanez JA, Remsberg CM, Forrest ML, Davies NM. Clinical toxicities of nanocarrier systems. Adv Drug Deliv Rev 2008; 60: 929-38.

[119] Brown DM, Donaldson K, Borm PJ, *et al.* Calcium and ROS-mediated activation of transcription factors and TNF-alpha cytokine gene expression in macrophages exposed to ultrafine particles. American J Physiol-Lung Cellular Mol Physiol 2004; 286: L344-53.

[120] Moss OR, Wong VA. When nanoparticles get in the way: Impact of projected area on *in vivo* and *in vitro* macrophage function. Inhalation Toxicology 2006; 18: 711-6.

[121] Unfried K, Albrecht C, Klotz LO, Von Mikecz A, Grether-Beck S, Schins RPF. Cellular responses to nanoparticles: target structures and mechanisms. Nanotoxicology 2007; 1: 52-71.

[122] Dobrovolskaia MA, McNeil SE. Immunological properties of Engineered nanomaterials. Nat Nanotechnol 2007; 2: 469-78.

[123] Shubayev VI, Pisanic TR, Jin SH. Magnetic nanoparticles for theragnostics. Advanced Drug Delivery Reviews 2009; 61: 467-77.

[124] Chellat F, Merhi Y, Moreau A, Yahia L. Therapeutic potential of nanoparticulate systems for macrophage targeting. Biomaterials 2005; 26: 7260-75.

[125] Moore A, Weissleder R, Bogdanov A. Uptake of dextran-coated monocrystalline iron oxides in tumor cells and macrophages. J Magn Reson Imaging 1997; 7: 1140-5.

[126] Storm G, Belliot SO, Daemen T, Lasic DD. Surface modification of nanoparticles to oppose uptake by the mononuclear phagocyte system. Advanced Drug Delivery Reviews 1995; 17: 31-48.

[127] Yamazaki M, Ito T. Deformation and instability in membrane structure of phospholipid vesicles caused by osmophobic association: mechanical stress model for the mechanism of poly(ethylene glycol)-induced membrane fusion. Biochemistry 1990; 29: 1309-14.

[128] Zhang Y, Kohler N, Zhang MQ. Surface modification of superparamagnetic magnetite nanoparticles and their intracellular uptake. Biomaterials 2002; 23: 1553-61.

[129] Sosnovik DE, Weissleder R. Emerging concepts in molecular MRI. Curr Opin Biotechnol 2007; 18: 4-10.

[130] Weber R, Wegener S, Ramos-Cabrer P, Wiedermann D, Hoehn M. MRI detection of macrophage activity after new experimental stroke in rate: New indicators for late appearance of vascular degradation? Magn Reson Med 2005; 54: 59-66.

[131] Brochet B, Deloire MSA, Touil T, *et al.* Early macrophage MRI of inflammatory lesions predicts lesion severity and disease development in relapsing EAE. Neuroimage 2006; 32: 266-74.

[132]   Gorantla S, Dou HY, Boska M, *et al*. Quantitative magnetic resonance and SPECT imaging for macrophage tissue migration and nanoformulated drug delivery. J Leukoc Biol 2006; 80: 1165-74.

[133]   Jander S, Schroeter M, Saleh A. Imaging inflammation in acute brain ischemia. Stroke 2007; 38: 642-5.

[134]   Renwick LC, Brown D, Clouter A, Donaldson K. Increased inflammation and altered macrophage chemotactic responses caused by two ultrafine particle types. Occup Environ Med 2004; 61: 442-7.

[135]   Fang C, Shi B, Pei YY, Hong MH, Wu J, Chen HZ. *In vivo* tumor targeting of tumor necrosis factor-alpha-loaded stealth nanoparticles: Effect of MePEG molecular weight and particle size. Eur J Pharm Sci 2006; 27: 27-36.

[136]   Xia T, Kovochich M, Brant J, *et al*. Comparison of the abilities of ambient and manufactured nanoparticles to induce cellular toxicity according to an oxidative stress paradigm. Nano Lett 2006; 6: 1794-807.

[137]   Oberdorster G, Maynard A, Donaldson K, *et al*. Principles for characterizing the potential human health effects from exposure to nanomaterials: Elements of a screening strategy. Part Fibre Toxicol 2005; 2: 1-35.

[138]   Albrecht C, Borm PJA, Unfried K. Signal transduction pathways relevant for neoplastic effects of fibrous and non-fibrous particles. Mutation Research-Fundamental and Molecular Mechanisms of Mutagenesis 2004; 553: 23-35.

[139]   Gurr JR, Wang ASS, Chen CH, Jan KY. Ultrafine titanium dioxide particles in the absence of photoactivation can induce oxidative damage to human bronchial epithelial cells. Toxicology 2005; 213: 66-73.

[140]   Borm PJA, Robbins D, Haubold S, *et al*. The potential risks of nanomaterials: a review carried out for ECETOC, Part Fibre Toxicol 2006; 3: 1-35.

[141]   Medina C, Santos-Martinez MJ, Radomski A, Corrigan OI, Radomski MW. Nanoparticles: Pharmacological and toxicological significance. Br J Pharmacol 2007; 150: 552-8.

[142]   Page-McCaw A, Ewald AJ, Werb Z. Matrix metalloproteinases and the regulation of tissue remodeling. Nat Rev Mol Cell Biol 2007; 8: 221-33.

[143]   Nelson KK, Melendez JA. Mitochondrial redox control of matrix metalloproteinases. Free Radic Biol Med 2004; 37: 768-84.

[144]   Liu KJ, Rosenberg GA. Matrix metalloproteinases and free radicals in cerebral ischemia. Free Radic Biol Med 2005; 39: 71-80.

[145]   Rosenberg GA. Matrix metalloproteinases in neuroinflammation. Glia 2002; 39: 279-91.

[146]   Kieseier BC, Hartung HP, Wiendl H. Immune circuitry in the peripheral nervous system. Curr Opin Neurol 2006; 19: 437-45.

[147]   Myers RR, Campana WM, Shubayev VI. The role of neuroinflammation in neuropathic pain: mechanisms and therapeutic targets. Drug Discov Today 2006; 11: 8-20.

[148]   Shubayev VI, Angert M, Dolkas J, Campana WM, Palenscar K, Myers RR. TNF alpha-induced MMP-9 promotes macrophage recruitment into injured peripheral nerve. Mol Cell Neurosci 2006; 31: 407-15.

[149]   Gurzau ES, Neagu C, Gurzau AE. Essential metals - case study on iron. Ecotoxicol Environ Saf 2003; 56: 190-200.

[150]   Bush AI. Metals and neuroscience. Curr Opin Chem Biol 2000; 4: 184-91.

[151]   Ke Y, Qian ZM. Iron misregulation in the brain: a primary cause of neurodegenerative disorders. Lancet Neurology 2003; 2: 246-53.

[152]   Doraiswamy PM, Finefrock AE. Metals in our minds: Therapeutic implications for neurodegenerative disorders. Lancet Neurology 2004; 3: 431-4.

[153]   Hautot D, Pankhurst QA, Khan N, Dobson J. Preliminary evaluation of nanoscale biogenic magnetite in Alzheimer's disease brain tissue. Proceedings of the Royal Society B-Biological Sciences 2003; 270: S62-4.

[154]   Hautot D, Pankhurst QA, Morris CM, Curtis A, Burn J, Dobson J. Preliminary observation of elevated levels of nanocrystalline iron oxide in the basal ganglia of neuroferritinopathy patients. Biochim Biophys Acta Mol Basis Dis 2007; 1772: 21-5.

[155]   Dobson J. Gene therapy progress and prospects: Magnetic nanoparticle-based gene delivery. Gene Therapy 2006; 13: 283-7.

[156]   Kim JS, Yoon TJ, Kim BG, *et al*. Toxicity and tissue distribution of magnetic nanoparticles in mice. Toxicol Sci 2006; 89: 338-47.

[157]   Liu G, Men P, Harris PLR, Rolston RK, Perry G, Smith MA. Nanoparticle iron chelators: A new therapeutic approach in Alzheimer disease and other neurologic disorders associated with trace metal imbalance. Neuroscience Letters 2006; 406: 189-93.

[158]   Green-Sadan T, Kuttner Y, Lublin-Tennenbaum T, *et al*. Glial cell line-derived neurotrophic factor-conjugated nanoparticles suppress acquisition of cocaine self-administration in rats. Exp Neurol 2005; 194: 97-105.

[159]   Powers KW, Palazuelos M, Moudgil BM, Roberts SM. Characterization of the size, shape, and state of dispersion of nanoparticles for toxicological studies. Nanotoxicology 2007; 1: 42-51.

[160] Teeguarden JG, Hinderliter PM, Orr G, Thrall BD, Pounds JG. Particokinetics *in vitro*: Dosimetry considerations for *in vitro* nanoparticle toxicity assessments. Toxicol Sci 2007; 95: 300-12.

[161] Warheit DB, Borm PJA, Hennes C, Lademann J. Testing strategies to establish the safety of nanomaterials: Conclusions of an ECETOC workshop. Inhalation Toxicology 2007; 19: 631-43.

[162] Worle-Knirsch JM, Pulskamp K, Krug HF. Oops they did it again! Carbon nanotubes hoax scientists in viability assays. Nano Lett 2006; 6: 1261-8.

[163] Barthel A, Klotz LO. Phosphoinositide 3-kinase signaling in the cellular response to oxidative stress. Biol Chem 2005; 386: 207-16.

[164] Muldoon LL, Sandor M, Pinkston KE, Neuwelt EA. Imaging, distribution, and toxicity of superparamagnetic iron oxide magnetic resonance nanoparticles in the rat brain and intracerebral tumor. Neurosurgery 2005; 57: 785-96.

[165] Dunning MD, Lakatos A, Loizou L, *et al.* Superparamagnetic iron oxide-labeled Schwann cells and olfactory ensheathing cells can be traced *in vivo* by magnetic resonance imaging and retain functional properties after transplantation into the CNS. Journal of Neuroscience 2004; 24: 9799-810.

[166] Gajdosikova A, Gajdosik A, Koneracka M, *et al.* Acute toxicity of magnetic nanoparticles in mice. Neuroendocrinology Letters 2006; 27: 96-9.

[167] Kante B, Couvreur P, Duboiskrack G, *et al.* Toxicity of polyalkylcyanoacrylate nanoparticles 1: Free nanoparticles. J Pharm Sci 1982; 71: 786-90.

CHAPTER 5

Hazards of TiO$_2$ and Amorphous SiO$_2$ Nanoparticles

Lucas Reijnders[*]

IBED, University of Amsterdam, Nieuwe Achtergracht 166, 1018 WV Amsterdam, The Netherlands

Abstract: TiO$_2$ and amorphous SiO$_2$ nanoparticles have been described as 'safe', 'non-toxic' and 'environment friendly' in scientific literature. However, though toxicity data are far from complete, there is evidence that these nanoparticles are hazardous. TiO$_2$ nanoparticles have been found hazardous to humans on inhalation, ingestion and dermal exposure. Ecotoxicity at levels of TiO$_2$ nanoparticles which are expected in the environment has also been found. Amorphous SiO$_2$ nanoparticles appear to be hazardous to humans on inhalation and ingestion and there is some evidence for ecotoxicity of amorphous SiO$_2$ nanoparticles. A main, though not the only, mechanism underlying the hazards of SiO$_2$ and TiO$_2$ nanoparticles may be the generation of reactive oxygen species. In view of the lack of scientific data pertinent to quantification of hazard and risk, a precautionary approach to production and usage of SiO$_2$ and TiO$_2$ nanoparticles has been advocated. Options for hazard reduction, such as coatings for TiO$_2$ nanoparticles, functionalization for amorphous SiO$_2$ nanoparticles and binding of nanoparticles to substrates, and risk reduction, including containment and membrane filtration, are discussed.

Keywords: Nanoparticle, Silica, Titania, TiO$_2$, SiO$_2$, Hazard, Toxicity, Ecotoxicity, Safety, Reactive oxygen species, Hazard reduction, Risk reduction, Precautionary principle.

TIO$_2$ AND AMORPHOUS SIO$_2$ NANOPARTICLES AND THEIR APPLICATIONS

Nanoparticles have been defined as particles $<$ 100 nm (nanometer) in at least one dimension. Nanoparticles have been categorized in manufactured and non-manufactured or engineered and non-engineered [1, 2]. Non-manufactured or non-engineered nanoparticles emerge as unintended by-product of human activities. For instance, burning diesel fuel tends to generate as unintended by-product soot nanoparticles. Manufactured or engineered nanoparticles are intended products. The latter does not preclude their unintended release, for instance in waste water as a result of cleaning operations, or due to accidental spillage.

Among the manufactured or engineered nanoparticles which are currently produced and used, amorphous SiO$_2$ (silica) and TiO$_2$ (titania) are among the most important [3, 4]. Both have a considerable variety of applications including food, pharmaceuticals, cosmetics, toothpastes, sunscreens, rubber products such as tyres, toners, paints, solar cells and 'self-cleaning' glass and ceramics [3, 4]. Many other applications, including nanocomposites with polymers and remediation technologies, are under development. Most of these applications pertain to nanoparticles $<$ 100 nm in all three dimensions, but there is also limited usage of TiO$_2$ nanoparticles with a length $>$ 100 nm (nanotubes, nanorods, nanofibres).

To a large extent, amorphous SiO$_2$ and TiO$_2$ nanoparticles are produced in factories and subsequently shipped to companies which apply these particles. However, there is also '*in situ*' production of SiO$_2$ and TiO$_2$ nanoparticles. In the latter case the nanoparticles are synthesized as part of the final product. An example of the latter is the production of 'self-cleaning' glass in factories, in which there is '*in situ*' production of TiO$_2$ particles by vapour deposition. The '*in situ*' and '*ex situ*' production strategies for nanoparticulate products differ in their potential for workplace exposure and releases into the environment, the former tends to give rise to lower workplace exposure and environmental releases into air [1].

Some of the applications of TiO$_2$ and SiO$_2$ nanoparticles are dispersive, which means that nanoparticles are

---

[*]Address correspondence to Lucas Reijnders: Institute for Biodiversity and Ecosystem Dynamics, University of Amsterdam, Nieuwe Achtergracht 166, 1018 WV Amsterdam, The Netherlands; Tel: +31-20-5256206; Fax: +31-20-5256269 E-mail: l.reijnders@uva.nl

Haseeb Ahmad Khan and Ibrahim Abdulwahid Arif (Eds)

automatically set free. Examples are the application of TiO_2 particles in sunscreens and the use of SiO_2 and TiO_2 nanoparticles as glidants in powders for the food, cosmetics and pharmaceutical industries [5, 6]. These dispersive applications lead to human exposure (for instance by ingestion of food or pharmaceuticals or dermal application of cosmetics) as well as releases into the environment. In context of the latter, the fate of SiO_2 and TiO2 nanoparticles in facilities for waste water treatment is of interest. TiO_2 nanoparticles have been found to be only partly removed by waste water treatment [7, 8]. Jarvie *et al.* [9] have shown that, at relatively high concentrations, SiO_2 coated Fe nanoparticles are not removed from wastewater by primary treatment. However, the same nanoparticles with a coating of the non-ionic detergent, Tween 20, at similar concentrations causes flocculation and their removal from the waste water.

Other applications of nanoparticles are non-dispersive. In such cases the nanoparticles are immobilized in products. Examples are 'self cleaning' ceramics and windows, a variety of paints including TiO_2 or SiO_2 nanoparticles and 'superhydrophobic' materials with surfaces consisting of SiO_2 nanoparticles. Though such applications have been called 'inherently safe' [10], the possibility exists that nanoparticles may be released. Kaegi *et al.* [11] provide evidence that SiO_2 nanoparticles used in exterior wall paints are detached from new and aged paints by natural weathering. Hsu and Chein [12] have found a substantial release of TiO_2 from synthetic polymers coated with TiO_2 nanoparticles. There may even be intentional loss of nanoparticles when these are bound to, or incorporated, in polymeric products. Chalking paints for exterior decorative applications [13] are a case in point. These paints use the photochemical activity of TiO_2 to degrade the (organic) top layer of the paint. During rain storms the degraded top layer washes off [13]. Moreover, the wear, tear and maintenance may give rise to the loss of nanoparticles from such products. Tribological studies on SiO_2/acrylate nanocomposites show that friction leads to the gradual loss of SiO_2 nanoparticles [14]. In case of wear and tear it may well be that the nanoparticles are part of larger particles [15].

ARE TIO_2 AND AMORPHOUS SIO_2 NANOPARTICLES HAZARDOUS?

In scientific literature TiO_2 nanoparticles have been described as safe [16, 17] environment-friendly [18, 19], non-toxic [20-30] and environmentally benign [31]. SiO_2 nanoparticles have also been regarded as environmentally safe [32]. Research into the safety, toxicity and environmental aspects of SiO_2 and TiO_2 nanoparticles is still in its early stages. For this reason the data are far from complete. This holds proper for the unmodified TiO_2 and amorphous SiO_2 nanoparticles, but even more for these nanoparticles with surface modification (*e.g.,* by coating or functionalization). Such surface modifications may have a significant effect on aggregation, uptake and toxicity [33].

Still, there is substantial evidence that TiO_2 nanoparticles can be hazardous to humans. Hazard refers to the potential to harm. When there is exposure to hazardous particles this may lead to risk (chance that harm is done). For risk to occur, exposure to TiO_2 nanoparticles and other particles with a similar effect should exceed a certain level. Due to lack of data this level is highly uncertain in the case of ingestion or dermal exposure. However in the case of inhalation there is convincing evidence that in urban environments background exposure of humans to particulates is such that risk of lung and cardiovascular disease occurs, and that added exposure to TiO_2 nanoparticles may increase such risk [34-37]. Persons with asthma or chronic obstructive pulmonary disease may be more at risk than healthy people, because there is less clearance of nanoparticles from their lungs [38].

Inhaled TiO_2 nanoparticles have been found to be hazardous to lung tissue and, dependent on size, to the cardiovascular system, with effects including inflammation, fibrosis and damage to DNA or genotoxicity [4, 5, 34, 35, 39-53]. Rossi *et al.* [54] have found that inhaled SiO_2 coated TiO_2 nanoparticles are more potent in causing inflammation of the lungs than non-coated TiO_2 nanoparticles. TiO_2 nanoparticles, whether or not coated with amorphous silica, have furthermore been shown to be cytotoxic to, and to activate, antigen presenting cells, such as bone marrow derived dendritic cells and macrophages [55], giving rise to unintended immune responses.

The finding of genotoxic effects by TiO_2 nanoparticles may well be linked to lung cancer associated with the inhalation of TiO_2 nanoparticles by rodents, which in turn varies depending on antioxidant protection which is different in rodent species [39]. There is as yet no epidemiological study which for the general human population clarifies antioxidant protection in human lungs, and the susceptibility of humans to TiO$_2$ nanoparticle induced lung cancer. It may be noted though, that TiO$_2$ has been classified as a possible carcinogen for humans [56].

Hazard to the cardiovascular system after inhalation is probably linked to translocation from the lungs [35, 52, 57]. Such translocation may also negatively affect other organs when large numbers of TiO$_2$ particles are translocated and poorly cleared from the body [2, 35, 58, 59]. In the context of the latter, it may be noted that spherical nanoparticles can be rapidly cleared from the blood circulations when their size < 5.5 nm [59]. Studies on other mineral nanoparticles suggest that the efficiency in translocation from the lungs to the blood circulation may be up to a few percent, but that the efficiency may be substantially larger for very small nanoparticles (diameter 1-4 nm) [60, 61]. Nanoparticles deposited in the cardiovascular system may give rise to myocardial infarction, augmented ischemia-reperfusion injury, arrhythmias and altered heart rate variability and to disruption of microvascular reactivity and nitric oxide signaling [62, 63]. TiO$_2$ nanoparticles which have entered the bloodstream may partly be deposited in organs such as liver and kidney [2].

No peer reviewed report on toxicological testing of TiO$_2$ nano-tubes, -rods or -fibres has been found. It should be noted, however, that concerns have been raised about mineral nanoparticles with a length exceeding 10 micrometer, because macrophages which remove particles from the lungs can not completely engulf such particles, a phenomenon called 'frustrated phagocytosis' [64]. This may be conducive to pathological responses comparable to the responses to asbestos [64]. Whether TiO$_2$ nano-tubes, -rods or -fibres would also be able to cause mesothelioma, like asbestos does, would seem to be dependent on the ability of particles of sufficient length to translocate to the pleura, to be retained by pleural tissues and cause carcinogenic effects in pleural tissue [65].

There is evidence that TiO$_2$ nanoparticles can be translocated from the nasal area to the central nervous system *via* the olfactory nerve and bulb, thus posing a hazard of inhaled nanoparticles to the central nervous system, including enhanced inflammation when large number of TiO$_2$ particles are translocated [46, 47, 52, 66, 67]. Ingested titania nanoparticles may be hazardous too, leading to inflammation of the intestines and may have genotoxic effects [4, 5, 68]. There is furthermore evidence that TiO$_2$ particles may be translocated from the intestines and deposited in organs such as heart, liver and spleen and induce inflammation and genotoxicity when large numbers of TiO$_2$ nanoparticles are translocated and not rapidly cleared from the body [4, 69, 70].

The actual hazard of TiO$_2$ nanoparticles to humans after inhalation or ingestion is dependent on number, diameter, surface area, surface charge, surface free energy, aggregation/agglomeration and crystal structure [5, 34, 44, 52, 59, 70-75]. As to the latter, three types of TiO$_2$ crystals exist: rutile, anatase and brookite. The former two dominate in actual applications of TiO$_2$ nanoparticles, be it apart or in mixtures. Of these, *ceteris paribus,* anatase tends to be more hazardous than rutile [74, 76]. Anatase TiO$_2$ nanoparticles tend to induce cell necrosis and rutile nanoparticles apoptosis [74]. It has been argued that changes in size of TiO$_2$ nanoparicles may be associated with non-linear changes in toxicity. Especially TiO$_2$ nanoparticles < 30 nm might well have unique properties which give rise to non-linear toxic responses [73].

Dermal exposure to TiO$_2$ nanoparticles may lead their entry into the living part of the skin, which in turn may give rise to inflammation and genotoxicity. Dermal exposure to TiO$_2$ nanoparticles may be caused through skin penetration when the skin is broken and when TiO$_2$ nanoparticles remained on the intact skin for a very long time [2, 77, 78]. The ability to penetrate through the skin may be linked to oxidative damage of the skin caused by TiO$_2$ nanoparticles [78]. After penetration through the skin TiO$_2$ nanoparticles may aggravate atopic dermatitis [79].

There is evidence that TiO$_2$ nanoparticles can exhibit ecotoxicity in water, enhanced by ultraviolet irradiation, and may negatively affect fish, the bivalve Mytilus, crustaceans and a variety of unicellular organisms including a number of bacteria and algae [6, 36, 37, 80-90]. Size, surface area and aggregation state have emerged as important determinants of nano TiO$_2$ ecotoxicity. Negative effects on aquatic microbial communities may occur at TiO_2 nanoparticle levels which are to be expected in the aquatic environment of industrialized countries, especially when heavily impacted by sewage treatment effluents [82, 91]. Direct toxicity of TiO_2 to fish is unlikely to be a major hazard [88], but TiO_2 nanoparticles may enhance the bioaccumulation of cadmium and arsenate in fish [92-94]. TiO_2 nanoparticles can also be ecotoxic in soils and sediments, *e.g.,* negatively affecting unicellular organisms, nematodes and worms [89, 95, 96]. It has furthermore been pointed out that TiO_2 nanoparticles may affect the relative amounts of phosphate in sediment and surface water [97]. At high levels of solar irradiation, eutrophication or nutrification of surface water might be increased, which in turn might increase the risk of toxic algal blooms.

Amorphous SiO_2 nanoparticles can be hazardous to humans when inhaled, with cytotoxicity strongly dependent on size and surface area of the nanoparticles [98-105]. Inhaled SiO_2 nanoparticles can increase the risk of pulmonary and cardiovascular disease in contexts where there is already a large environmental exposure to non-manufactured small mineral particles, as is common in urban areas [1, 34-37]. People with asthma or chronic obstructive pulmonary disease may be more at risk than healthy people, because there is less clearance of nanoparticles from their lungs [38]. Major determinants of SiO_2 nanoparticle hazard apart from size and surface area would seem: surface charge, surface free energy and agglomeration [73, 75]. It may well be that changes in size of amorphous SiO_2 nanoparticles can give rise to non-linear changes in toxicity, especially when amorphous SiO_2 nanoparticles are < 30 nm [73]. Roiter *et al.* [106] have demonstrated that there is non linearity in interaction between spherical amorphous SiO_2 nanoparticles below and above a 22 nm diameter. When the diameter is > 22 nm smooth silica nanoparticles were completely enveloped by a lipid membrane, whereas at a diameter <22 nm the lipid membrane underwent structural rearrangements by forming holes or pores [106].

Ingested SiO_2 nanoparticles may have cytoxic effects on tissues of the intestines, especially after accumulation in inflamed intestinal tissue [107] and may give rise to genotoxicity in intestinal cells [108]. Amorphous silica nanoparticles may partly pass from the lung, and probably also from the intestines, into the cardiovascular system. When spherical SiO_2 nanoparticles are < 5.5 nm they may be relatively rapidly cleared (majority in 4 hours) from circulating blood by the kidneys, unlike larger SiO_2 nanoparticles [59].

Following translocation from the lungs or intestines, SiO_2 nanoparticles may partly be distributed to organs such as liver, spleen, kidney and heart [57, 105, 109, 110] and may be hazardous to those organs. A similar property probably applies to siloxane coated silica nanoparticles [111]. There is empirical evidence that amorphous silica nanoparticles may be hepatotoxic [105, 110], toxic to embryonic kidney cells [112] and human endothelial cells [102] and may inhibit stem cell differentiation [113]. Amorphous SiO_2 nanoparticles may also exhibit ecotoxicity. Fujiwara *et al.* [114] and van Hoecke *et al.* [115] found toxicity of amorphous SiO_2 nanoparticles to the green algae Chlorella and Pseudokirchineriella subcapita at exposures which might occur in the environment. Canesi *et al.* [87] found immunotoxicty of SiO_2 nanoparticles to the bivalve Mytilus.

A main molecular mechanism of cytotoxicity in case of amorphous SiO_2 and TiO_2 nanoparticles in the absence of light, may be: oxidative damage linked to reactive oxygen species, whereas TiO_2 particles exposed to light and/or UV radiation may also damage cells due to photocatalytically enhanced oxidation [35, 43, 62, 63, 66, 68, 71, 78, 83, 87, 101, 105, 112, 116-124]. Reactive oxygen species in turn may trigger cytokines and upregulate several other cellular components such as p53 and Bax protein, and may negatively affect a variety of cellular components, including DNA [43, 50, 62, 66, 78, 105, 118, 120]. Studies so far suggest that enhanced oxidation by nanoparticles can be linked to structure (*e.g.,* crystallinity) and surface area [72, 125, 126].

However, enhanced oxidation may not be the only factor in causing toxicity. The relatively strong inflammatory effect of SiO_2-coated TiO_2 nanoparticles in rodent lungs found by Rossi *et al.* [54] is unlikely to be exclusively linked to enhanced oxidation. Also it seems unlikely that the necrotic effect of anatase can solely be explained on the basis of enhanced oxidation [74]. There is evidence that cytotoxicity of TiO_2 may be partly linked to the adsorption of proteins and Ca^{2+} [127]. Another possibility which has been raised is that nanoparticles may negatively affect DNA by their impact on signaling molecules such as ATP [128].

Furthermore, the possibility has been raised that nano-TiO_2 in the presence of light may give rise to protein tyrosine nitration, which may lead to photosensitized damage to tissues [129].

APPLICATION OF THE PRECAUTIONARY PRINCIPLE

As pointed out before, toxicity data regarding TiO_2 and amorphous SiO_2 nanoparticles are incomplete. And data about the releases of nanoparticles and levels of these nanoparticles in the environment are very limited. The limited availability of toxicity data makes it impossible to establish limit values in the way commonly used for substances with sufficient data-availability, such as lead, chlorodioxins or SO_2, for the workplace or the environment. This had led some authors to call for application of the precautionary principle [52, 130-133]. The precautionary principle aims at facilitating decision making under uncertainty, when negative effects have been identified but can, as yet, not be quantified. For instance, Helland *et al.* [130] have stressed the importance of eliminating problems at the source, which would mean a focus on hazard reduction, which is discussed in the next section. Oberdörster [52] has suggested 'appropriate precautionary measures/regulations and practicing best industrial hygiene to avoid future horror scenarios from environmental or occupational exposures'. 'Best practices', as advocated by Oberdörster [52], would imply hazard and risk reduction. These will be discussed in the two following sections. European trade unions and environmental non governmental organizations have called for regulations which would prevent exposure to, and emissions of, nanoparticles when data about their hazard and risk are insufficient [133]. As data about hazard and risk about TiO_2 and amorphous SiO_2 nanoparticles are far from complete, living up to this call would lead to elimination of dispersive uses of these particles, such as in cosmetic, medicines and food.

HAZARD REDUCTION

Firstly, hazard reduction may focus on alternatives which are inherently safer [36, 130]. In this context one might for instance consider the substitution of glidants based on TiO_2 or SiO_2 nanoparticles by stearates [5]. Secondly, one might consider reducing nanoparticle hazard. For instance, cytotoxicity of amorphous SiO_2 nanoparticles may be reduced by functionalization with amidopropytrimethoxysilane [134]. In the case of TiO_2 nanoparticles exposed to solar radiation, toxicity (and thereby hazard) is much enhanced by photocatalytic effects. To the extent that such effects are not essential to product performance (*e.g.,* in the case of applications in cosmetics such as sunscreens), photocatalytic effects can be much reduced by coatings. It has been argued that, irrespective of irradiation, the toxicity of TiO_2 nanoparticles can be reduced by apatite coatings [135]. Though it seems likely that such coatings can reduce photocatalytic activity, their impact on hazard in the absence of light would seem a matter of uncertainty. Oxidative stress, cytotoxic effects (including apoptosis), and aberrant immune responses to nanoparticles with an apatite surface have been reported [136, 137]. Commonly one finds silica coatings on TiO_2 nanoparticles used in sunscreens and other cosmetics [1, 6]. Such coatings decrease photocatalytic activity, but tests on the photocatalytic activity of silica-coated TiO_2 nanoparticles marketed by industry did show that actual reduction of photocatalytic activity by coating exhibited substantial variation [138]. Also one should note that, according to a study of Rossi *et al.* [54] silica-coated TiO_2 nanoparticles can be more potent in causing inflammation of the lungs than non-coated TiO_2 nanoparticles.

Biodegradable and biocompatible polymeric coatings have also been advocated to reduce nanoparticle toxicity [59]. The benefit thereof is contested as such coatings may degrade, which leads to exposure of the inorganic surface, inflammatory and immune responses may be activated [59]. Also, on entering blood circulation, nanoparticle circulation times of TiO_2 nanoparticles with an organic coating can be increased and accumulation in the reticuloendothelial system may occur [59]. The latter includes liver, spleen, lymph nodes and bone marrow.

The hazard of both TiO_2 and amorphous SiO_2 nanoparticles may be minimized by strong binding to substrates and product designs which only allow for the release of relatively large particles [36]. To the extent that TiO_2 nanoparticles are bound to substrates and products are or can be irradiated, it has been argued that to limit hazard binding to inorganic substrates is to be preferred to linkage to organic substrates, as the latter can be degraded by the photocatalytic activity of TiO_2 nanoparticles [84].

RISK REDUCTION

Both hazard reduction and reduction of exposure may contribute to risk reduction, which would be important in the best practices advocated by Oberdörster [52]. In the production of nanoparticles there may be substantial exposures of those working in the vicinity linked to the opening of hatches and doors of production equipment [139], and near equipment for filling, bagging and filtering [3]. Cleaning production equipment may also be a source of significant worker exposure [1]. In laboratory settings the use of fume hoods may still lead to considerable worker exposure on handling dry powders consisting of nanoparticles or on sonication [140, 141]. Respiratory protection of workers may reduce exposure, but it should be noted that state-of-the-art respiratory protection equipment may let through 1-10% of the nanoparticles in the 15 nm-100 nm range [142]. Exposure to nanoparticles may be more strongly reduced by containment and remote control [1, 36]. As pointed out before, waste water treatment with current mainstream technologies leads only to limited removal of nanoparticles from waste water streams [7, 8, 91]. In principle higher removal efficiencies might be attained by techniques such as membrane filtration.

CONCLUSION

TiO_2 and amorphous SiO_2 nanoparticles have been described as 'safe', 'non-toxic' and 'environment friendly' in scientific literature. However, though toxicity data are far from complete, there is evidence that these nanoparticles are hazardous. TiO_2 nanoparticles have been found hazardous to humans on inhalation, ingestion and dermal exposure. Ecotoxicty at levels of TiO_2 nanoparticles which are expected in the environment has also been found. Amorphous SiO_2 nanoparticles appear to be hazardous to humans on inhalation and ingestion and there is some evidence for ecotoxicity of amorphous SiO_2 nanoparticles. A main, though not the only, mechanism underlying the hazards of TiO_2 and SiO_2 nanoparticles may be the generation of reactive oxygen species. In view of the lack of scientific data pertinent to hazard and risk, a precautionary approach to production and usage of SiO_2 and TiO_2 nanoparticles has been advocated. Proposals for operationalization of the precautionary approach have suggested source reduction and best practices, which in turn imply hazard and risk reduction. Several options for hazard reduction, such as coatings and functionalization for nanoparticles and binding of nanoparticles to substrates, and several options for risk reduction, including remote control and membrane filtration, have been briefly discussed.

REFERENCES

[1]   Reijnders L. Cleaner nanotechnology and hazard reduction of manufactured nanoparticles. J Clean Prod 2006; 14: 124-133.

[2]   Casals E, Vazquez-Campos S, Bastus NG, Puntes V. Distribution and potential toxicity of engineered inorganic nanoparticles and carbon nanostructures in biological systems. Trends Anal Chem 2008; 27: 672-683.

[3]   Hanai S, Kobayashi N, Rma M, *et al.* Risk assessment of manufactured nanomaterials -Titanium Dioxide (TiO_2) - Interim report issued on October 16 2009. National Institute of Advanced Industrial Science and Technology Tokyo 2009.

[4]   Trouiller B, Reliene R, Westbrook A, Solaimani P, Schiestl RH. Titanium dioxide nanoparticles induce DNA damage and genetic instability *in vivo* in mice. Cancer Res 2009; 69: 8784-8789.

[5]   Reijnders L. Biological effects of nanoparticles used as glidants in powders. Powder Technol 2007; 175: 143-145.

[6]   Reijnders L. Safety of nanoparticles in sunscreens. Househ and Personal Care Today 2009; 3: 16-17.

[7]   Limbach LK, Bereiter R, Müller E, Krebs R, Galli R, Stark WJ. Removal of nanoparticles in a model wastewater treatment plant: influence of agglomeration and surfactants on clearing efficiency. Environ Sci Technol 2008; 42: 5828-5833.

[8]   Kiser MA, Westerhoff F, Benn T, Wang P, Perez-Rivera J, Hristovski K. Titanium nanomaterial removal and release from wastewater treatment plants. Environ Sci Technol 2009; 43: 6757-6763.

[9]   Jarvie HP, Al-Obaidi H, King SM, *et al.* Fate of silica nanoparticles in simulated primary waste water treatment. Environ Sci Technol 2009; 43: 8622-8628.

[10]   Horng J. Growing carbon nanotube on aluminum oxides. An inherently safe approach to environmental applications. Trans Inst Chem Engin. Process Safety Environ Prot 2007; 85: 332-339.

[11]   Kaegi R, Ulrich A, Sinnet B, *et al.* SiO₂ nanoparticle emission from exterior facades into the aquatic environment. Environ Pollut 2008; 156: 233-139.

[12]   Hsu L, Chein H. Evaluation of nanoparticle emission from TiO₂ nanopowder coating materials. J Nanopart Res 2007; 9: 157-163.

[13]   Fernando RH. Nancocmposite and nanostructured coatings: recent advancements. http://pubs.acs.org. 2009; DOI 10.1021/bk-2009-1008.ch001.

[14]   Devaprakasam D, Hatton PV, Möbus G, Inkson BJ. Effect of microstructure of nano- and micro-particle filled polymer composites on their tribo-mechanical performance. J Phys Conf Ser 2008; 126: 012057.

[15]   Koponen IK, Jensen KA, Schneider T. Sanding dust from nanoparticle-containing paints: physical characterization. J Phys Conf Seri 2009; 151: 012048.

[16]   Tompkins DT, Lawnicki BJ, Zeitner WA. Evaluation of photocatalysis for gas-phase air cleaning- part 1: process, technical and sizing considerations. ASHRAE Trans 2005; 111: 60-85.

[17]   Sakatani Y, Grosso D, Nicole L, Bossière C, Soler-Illia GJAA, Sanchez C. Optimized photocatalytic activity of grid like mesoporous TiO₂ films: effect of crystallinity, pore size distribution, and pore accessibility. J Mater Chem 2006; 18: 77-82.

[18]   Gu Z., Fujishama A, Sato O. Biomimetic titanium-dioxide film with structural color and extremely stable hydrophylicity. App Phys Lett 2004; 85: 5067-5069.

[19]   Mosurkal R, Samuelson IA, Smith KD, Westmoreland PR, Parmar VS, Yan F. Nanocomposites of TiO₂ and siloxane copolymers as environmentally safe flame retardant materials. J Macromol Sci A: Pure Appl Chem 2008; 45: 924-946.

[20]   Anderssson M, Österlund L, Ljungström S, Palmqvist A. Preparation of nanosize anatase and rutile TiO₂ by hydrothermal treatment of microemulsions and their activity for photocatalytic wet oxidation of phenol. J Phys Chem B 2002; 106: 10674-10679.

[21]   Li Z, Pengyi Z, Songzhe C. Influence of pretreatment of titanium substrate on long-term stability of TiO₂ film. Chin J Catal 2007; 28: 299-306.

[22]   Burlacov I, Jirkovsky J, Müller M, Heimann RB. Induction plasma-sprayed photocatalytically active titania coatings and their characterisation by micro-Raman spectroscopy. Surf Coat Technol 2006; 201: 255-264.

[23]   Hirano M and Matushima K. Photoactive and adsorptive niobium-doped anatase (TiO₂) nanoparticles: influence of hydrothermal conditions on their morphology, structure and properties. J Am Ceram Soc 2006; 89:110-117.

[24]   Messina PV, Morini MA, Sierra MB, Schulz PC. Mesoporous silica-titania composite materials. J Coll Interface Sci 2006; 300: 270-278.

[25]   Kubacka A, Serrano C, Ferrer M, Lunsdorf H, Bielecki P. Cerrada ML. High-performance dual action polymer-TiO₂ nanocomposite films *via* melting processing. Nano Lett 2007; 7: 2529-2534.

[26]   Liuxue Z, Xiluian W, Peng L, Zhixing S. Photocatalytic activity of anatase thin films coated cotton fibres prepared *via* a microwave assisted liquid phase deposition process. Surf Coat Technol 2007; 201: 7607-7614.

[27]   Allen BS, Edge M, Verran J, Stratton J, Maltby J, Bygott C. Photocatalytic titania based surfaces: environmental benefits. Polym Degrad Stab 2008; 93: 1632-1646.

[28]   Immoos CE, Jaoudi MD, Gu FK, Wang DL. Reactive nanoparticles in coatings. http://pubs.acs.org, 2009; doi 10.1021/bk-2009-1008.ch010.

[29]   Tung WS and Daoud WA. Effect of wettability and silicone surface modification on the self-cleaning functionalization of wool. J Appl Polym Sci 2009; 112: 235-243.

[30]   Wu D, Long M, Zhou J, *et al.* Synthesis and characterization of self-cleaning cotton fabric modified by TiO₂ through a facile approach. Surf Coat Technol 2009; 203: 3728-3733.

[31]   Liu YY, Qian LQ, Guo C, Jia X, Wang JW, Tang WH. Natural hydrophilic TiO₂/SiO₂ composite thin films deposited by radio frequency magnetron sputtering. J Alloy Compounds 2009; 479: 529-535.

[32]   Mizutani T, AraiK, Miyamoto M, Kimura Y. Application of silica-containing nanocomposite emulsion to wall paint: a new environmentally safe paint of high performance. Progr Org Coat 2007; 55: 276-283.

[33]   Bystrzejewska-Piotrowska G, Golimowski J, Urban PL. Nanoparticles: their potential toxicity, waste and environmental management. Waste Manag 2009; 29; 2587-2595.

[34]   Borm PJA and Beruba D. A tale of opportunities, uncertainties, and risks. NanoToday 2008; 3: 56-59.

[35]   Oberdörster G, Stone V, Donaldson K. Toxicology of nanoparticles: a historical perspective. Nanotoxicol 2007; 1: 2-25.

[36]   Reijnders L. Hazard reduction in nanotechnology. J Ind Ecol 2008; 12(3): 297-306.

[37] Reijnders L. Hazard reduction for the application of titania nanoparticles in environmental technology. J Hazard Materi 2008; 142: 440-445.

[38] Kreyling WG, Semmler-Behnke M, Möller W. Ultrafine particle lung-interactions: does size matter? J Aerosol Med 2006; 19: 74-83.

[39] Borm PJA, Schins RPF, Albrecht C. Inhaled particles and lung cancer. Part B. Paradigms and risk assessment. Int J Cancer 2004; 110: 3-14.

[40] NIOSH. Evaluation and health hazard and recommendations for the occupational exposure to titaniumdioxide. NIOSH Curr Intell Bull 2005.

[41] Brown SC, Kamal M, Nasreen N, *et al.* Influence of shape, adhesion and simulated lung mechanics on amorphous silica nanoparticle toxicity. Adv Powder Technol 2007; 18: 69-79.

[42] Liao C, Chiuang Y, Chio C. Assessing airborne titanium dioxide nanoparticle related exposure hazard at workplace. J Hazard Mater 2008; 162: 57-65.

[43] Long TC, Tajuba J, Sama P, *et al.* Nanosize titanium dioxide stimulates reactive oxygen species in brain microglias and damages neurons *in vitro.* Environ Health Persp 2007; 115: 1631-1637.

[44] Mueller NC, Nowack B. Exposure modeling of engineered nanoparticles in the environment. Environ Sci Technol 2008; 42: 4447-4453.

[45] Simon-Deckers A, Gouget B, Mayne-L'Hermite M, Herlin-Boime N, Raynaud C, Carrière M. *In vitro* investigation of oxide nanoparticle and carbon nanotube toxicity and intracellular accumulation in A549 human pneumocytes. Toxicol 2008; 253: 137-146.

[46] Wang J, Chen C, Liu Y, *et al.* Potential neurological lesion after nasal installation of TiO$_2$ nanoparticles in the anatase and rutile crystal phases. Toxicol Lett 2008; 183: 72-80.

[47] Wang J, Liu Y, Jiao F, *et al.* Time-dependent translocation and potential impairment on central nervous system by intranasally installed TiO$_2$ nanoparticles. Toxicol 2008; 254: 82-90.

[48] Falck GCM, Lindberg HK, Suhonen S, *et al..* Genotoxic effects of nanosized and fine TiO$_2$. Hum Exp Toxicol 2009; 28: 339-352.

[49] Kobayashi N, Naya M, Endoh S, Maru J, Yamamoto K, Nakanishi J. Comparative pulmonary toxicity study of nano-TiO$_2$ particles of different sizes and agglomeration in rats: different short- and long-term post-installation results. Toxicol 2009; 264: 110-118.

[50] Singh N, Manshian B, Jenkins GJS, *et al.* Nanogenotoxity: the DNA damaging potential of engineered nanomaterials. Biomater 2009; 30: 3891-3914.

[51] Barillet S, Simon-Deckers A, Herlin-Boime N, *et al.* Toxicological consequences of TiO$_2$, SiC nanoparticles and multi-walled carbon nanotubes exposure in several mammalian cell types: an *in vitro* study. J Nanopart Res 2010; 12: 61-73

[52] Oberdörster G. Safety assessment for nanotechnology and nanomedicine: concepts of nanotoxicology. J Internal Med 2010: 267: 89-105.

[53] Shi Y, Zhang J, Jiang M, Zhu L, Ton H, Lu B. Synergistic genotoxicity caused by low concentration of titanium dioxide nanoparticles and p, p'-DDT in human hepatocytes. Exp Mol Mutagen 2010; 51: 192-204.

[54] Rossi EM, Pylkänen L, Koivisto AJ, *et al.* Airway exposure to silica coated TiO$_2$ nanoparticles induces pulmonary neutrophilia in mice. Toxicol Sci 2010; 113: 422-433.

[55] Palomäki J, Karisola P, Pylkkänen L, Savolainen K, Alenius H. Engineered nanomaterials cause cyctotoxicity and activation on mouse antigen presenting cells. Toxicol 2010; 267: 125-131

[56] IARC Working Group. Carcinogenicity of carbon black, titanium dioxide, and talc. Lancet Oncol 2006; 7: 295-296.

[57] Oberdörster G, Oberdörster E, Oberdörster J Nanotoxicology: an emerging discipline evolving from studies of ultrafine particles. Environ Health Persp 2005; 113: 823-839.

[58] Semmler-Behnke M, Kreyling WG, Lipka J, *et al.* Biodistribution of 1.4 and 18 nm gold particles in rats. Small 2008; 4: 2108-2111.

[59] Choi HS, Liu W, Liu F, *et al.* Design conditions for tumor targeted nanoparticles. Nature Nanotechnology 2010; 5: 42-47.

[60] Kreyling WG, Semmler M, Erbe F, *et al.* Translocation of ultrafine insoluble iridium particles from lung epithelium to extrapulmonary organs is size dependent but very low. J Toxicol Environ Health A 2002; 65: 1513-1530.

[61] Semmler-Behnke M, Takenaka S, Fertsch S, *et al.* Efficient elimination of inhaled nanoparticles from the alveolar region. Evidence for interstitial uptake and subsequent re-entrainment onto airways epithelia. Environ Health Persp 2007; 115: 738-735.

[62] Nurkiewicz TR, Porter DW, Hubbs AF, *et al.* Pulmonary nanoparticle exposure disrupts systemic microvascular nitric oxide signaling. Toxicol Sci 2009; 110: 191-203.

[63] Leblanc AJ, Moseley AM, Chen BT, Frazer S, Castranova V, Nurkiewicz TR. Nanoparticle inhalation impairs coronary microvascular reactivity *via* a local reactive oxygen species-dependent mechanism. Cardiovasc Toxicol 201; 10: 651-663.

[64] Borm P, Castranova V. Toxicology of nanomaterials: permanent interactive learning. Part Fibre Toxicol 2009; 6: 28.

[65] Donaldson K, Poland CA. New insights in nanotubes. Nature Nanotechnol 2009; 4: 708-710.

[66] Long TC, Saleh N, Tilton RD, Lowry GV, Veronesi B. Titanium dioxide (P25) produces reactive oxygen species in immortalized brain microglia (BV2): implications for nanoparticle neurotoxicity. Environ Sci Technol 2006; 40: 4346-4352.

[67] Shin JA, Lee EJ, Seo SM, Kim HS, Kang JL, Park EM. Nanosized titanium dioxide enhanced inflammatory responses in the septic brain of mouse. Neuroscience 2010; 165: 445-454.

[68] Ma L, Liu J, Li N, *et al.* Oxidative stress in the brain of mice caused by translocated nanoparticulate TiO₂ delivered to the abdominal cavity. Biomater 2010; 31: 99-105.

[69] Janie PU, McCarthy DE, Florece AT. Titanium dioxide (rutile) particle uptake from the rat gastrointestinal tract and translocation to systemic organs after oral administration. Int J Pharmacol 1994; 1005: 157-168.

[70] Wang J, Zhou G, Chen C, *et al.* Acute toxicity and biodistribution of different sized titanium dioxide particles in mice after oral administration. Toxicol Lett 2007; 168: 176-185.

[71] Singh S, Shi T, Duffin R, *et al.* Endocytocis, oxidative stress and IL 8 expression in human lung epithelial cells upon treatment with fine and ultrafine TiO₂: role of the specific surface area and of surface methylation of the particles. Toxicol Appl Pharmacol 2007; 222: 141-151.

[72] Warheit DB, Webb TR, Reed KL, Frerichs S, Sayes CM. Pulmonary toxicity study in rats with three forms of ultrafine-TiO₂ particles: differential responses related to surface properties. Toxicol 2007; 230: 90-104.

[73] Auffan M, Rose J, Bottero J, Lowry GV, Jolivet J, Wiesner MA. Towards a definition of inorganic nanoparticles from an environmental, health and safety perspective. Nature Nanotechnol 2009; 4: 634-641.

[74] Braydich-Stolle LK, Schaeublin NM, Murdock RC, *et al.* Crystal structure mediates mode of cell-death in TiO₂ nanotoxicity. J Nanopart Res 2009; 11: 1361-1374.

[75] Jiang J, Oberdörster G, Biswas P. Characterization of size, surface charge, and agglomeration state of nanoparticle dispersions for toxicological studies. J Nanopart Res 2009; 11: 77-89.

[76] Sayes CM, Wahi R, Kurian PA, LiuV, West JL, Ausman KD, *et al.* Correlating nanoscale titania structure with toxicity and inflammatory response study with human dermal fibroblasts and human lung epithelial cells. Toxicol Sci 2006; 92: 174-185.

[77] Crosera M, Boivenzi M, Maina G, *et al.* Nanoparticle dermal absorption and toxicity: a review of the literature. International Arch Occup Environ Health 2009; 82: 1043-1055.

[78] Wu J, Liu W, Xue C, Zhou S, Lan F, Bi L, *et al.* Toxicity and penetration of TiO₂ nanoparticles in hairless mice and porcine skin after subchronic dermal exposure. Toxicol Lett 2009; 191: 1-8.

[79] Yanagisawa R, Takano H, Inoue K, *et al.* Titanium dioxide nanoparticles aggravate atopic dermatitis-like skin lesions in NC/Nga mice. Exp Biol Med 2009; 234: 314-322.

[80] Vevers WP and Jha AN. Genotoxic and cytotoxic potential of titanium dioxide (TiO₂) nanoparticles on fish cells *in vitro*. Ecotoxicol 2008; 17: 410-420.

[81] Wang J, Zhang X, Chen Y, Sommerfeld M, Hu Q. Toxicity assessment of manufactured nanomaterials using the unicellular green alga Chlamydomonas reinhardii. Chemosphere 2008; 73: 1121-1128.

[82] Battin TJ, Kammer FVS, Weilhatner A, Ottofuelling S, Hofman T. Nanostructured TiO₂: transport behaviour and effects on aquatic microbial communities under environmental conditions. Environ Sci Technol 2009; 41: 8098-8104.

[83] Farré M, Gajda-Schrantz K, Kantiani L, Barcelo D. Ecotoxity and analysis of nanomaterials in the aquatic environment. Anal Bioanal Chem 2009; 393: 81-95.

[84] Reijnders, L. The release of TiO₂ and SiO₂ nanoparticles form nanocomposites. Polym Degrad Stab 2009; 94: 873-876.

[85] Simon-Deckers A, Loo S, Mayne-l'Hermite, *et al.* Size- composition- and shape-dependent toxicological impact of metal oxide nanoparticles and carbon nanotubes toward bacteria. Environ Sci Technol 2009; 43: 8423-8429.

[86] Wiench K, Wohlleben W, Hisgen V, *et al.* Acute and chronic effects of nano- and non-nano-scale TiO₂ and Zn O particles on mobility and reproduction of the freshwater invertebrate Daphnia magna. Chemosphere 2009; 76: 1356-1365.

[87]   Canesi L, Ciacci C, Valotto D, Gallo G, Marcomini A, Pojan G. *In vitro* effects of suspensions of selected nanoparticles (C$_{60}$ fullerene, TiO$_2$, SiO$_2$) on Mytilus hemocytes. Aquat Toxicol 2010; 96: 151-158.

[88]   Johnston BD, Scown TM, Moger J, *et al.* Bioavailability of nanoscale metal oxides TiO$_2$, CeO$_2$ and ZnO to fish. Environ Sci Technol 2010; 44: 1144-1151.

[89]   Kahru A, Dubourguier H. From ecotoxicology to nanoecotoxicology. Toxicol 2010; 269: 105-119.

[90]   Zhu X, Chang Y, Chen Y. Toxicity and bioaccumulation of TiO$_2$ nanoparticle aggregates in Daphnia magna. Chemosphere 2010; 78: 209-215.

[91]   Gottschalk F, Sonderer T, Scholz RW, Nowack B. Modeled environmental concentrations of engineered nanomaterials (TiO$_2$, ZnO, Ag, CNT, fullerenes) for different regions. Environ Sci Technol 2009; 43: 9216-9222.

[92]   Sun H, Zhang Z, Niu Q, Chen Y, Crittenden JC. Enhanced accumulation of arsenate in carp in the presence of titanium dioxide nanoparticles. Water Air Soil Pollut 2007; 178: 245-254.

[93]   Sun H, Zaman X, Zhang Z, Chen Y, Crittenden JC. Influence of titanium dioxide nanopariicles on speciation and bioavailability of arsenite. Environ Pollut 2009; 157: 1165-1170.

[94]   Zhang X, Sun H, Zhang Z, Niu Q, Chen Y, Crittenden JC. Enhanced bioaccumulation of cadmium in carp in the presence of titanium dioxide nanoparticles. Chemosphere 2007; 67: 160-166.

[95]   Navarro E, Baun A, Behra R, *et al.* 2008. Environmental behavior and ecotoxicity of engineered nanoparticles to algae, plants and fungi. Ecotoxicol 2008; 17: 372-386.

[96]   Galloway T, Lewis C, Dolciotti I, Johnston B, Moger J, Regoli F. Sublethal toxicity of nano-titanium dioxide and carbon nanotubes in a sediment dwelling marine polychaete. Environ Pollut 2010; 158: 1748-1755.

[97]   Luo Z, Wang Z, Li Q, Pan Q, Yan C. Effects of titania nanoparticles on phosphorus fractions and its release in resuspended sediments under UV irradiation. J Hazard Mater 2010; 174: 477-483.

[98]   Chen M and Mikecz A von. Formation of nucleoplasmic protein aggregates impairs nuclear function in response to SiO$_2$ nanoparticles. Exp Cell Res 2005; 305: 51-62.

[99]   Chang J, Chang KLB, Hwang D, Kong Z. *In vitro* cytotoxicity of silica nanoparticles at high concentrations strongly depends on the metabolic activity type of the cell line. Environ Sci Technol 2007; 41: 2064-2068.

[100] Chen Z, Meng H, Xing G, *et al.* Age-related differences in pulmonary and cardiovascular responses to SiO$_2$ nanoparticle inhalation: nanotoxicity has susceptible population. Environ Sci Technol 2008; 42: 8985-8992.

[101] Eom H, Choi J. Oxidative stress of silica nanoparticles in human bronchial epithelial cell, Beas-2B. Toxicol *in vitro* 2009; 23: 1326-1332.

[102] Napierska D, Thomassen LCJ, Rabolli V, *et al.* Size-dependent cytotoxicity of monodisperse silica nanoparticles in human endothelial cells. Small 2009; 5: 846-853.

[103] Waters KM, Masiello LM, Zangar RC, *et al.* Macrophage responses to silica nanoparticles are highly conserved across particle sizes. Toxicol Sci 2009;107: 553-569.

[104] Yu KO, Grabinski CM, Schrand AM, *et al.* Toxicity of amorphous silica nanoparticles in mouse keratinocytes. J Nanoparticle Res 2009; 11: 15-24.

[105] Ye Y, Liu J, Xu J, Sun L, Chen M, Lan M. Nano-SiO$_2$ induces apoptosis *via* activation of p 53 and Bax mediated by oxidative stress in human cell line. Toxicol *in vitro* 2010; 24: 751-758

[106] Roiter Y, Ornatska M, Rammohan AR, Balakrishnan B, Heine DR, Minko S. Interaction of lipid membrane with nanostructural surfaces. Langmuir 2009; 25: 6287-6299.

[107] Moulari B, Pertuit D, Pellequer Y, Lamprecht A. The targeting of surface modified silica nanoparticles to inflamed tissue in experimental colitis. Biomater 2008; 29: 4554-4560.

[108] Gerloff K, Albrecht C, Boots AW, Forster I, Schins RPF. Cytotoxicity and oxidative DNA damage by nanoparticles in human intestinal Caco-2 cells. Nanotoxicol 2009; 3: 355-364.

[109] Cho M, Cho W, Choi M, *et al.* The impact of size on tissue distribution and the elimination by single intravenous injection of silica nanoparticles. Toxicol Lett 2009; 189: 177-183.

[110] Nishimori H, Kondoh M, Isoda K, Tsunoda S, Tsustumi Y, Yagi K. Silica nanoparticles as hepatotoxicants. Eur J Pharm Biopharm 2009; 72: 496-501.

[111] Brindell JR. Nanotechnology and the dilemmas facing business and government. Florida Bar J 2009; July/August: 73-77.

[112] Wang F, Gao F, Lan M, Yuan H, Huang Y, Liu J. Oxidative stress contributes to silica nanoparticle-induced cytotoxicity in human embryonic kidney cells. Toxicol *in vitro* 2009; 23: 808-815.

[113] Park MVDZ, Annema W, Salvati A, *et al.* In vitro developmental toxicity test detects inhibition of stem cell differentiation by silica nanoparticles. Toxicol Appl Pharmacol 2009; 240: 108-116.

[114] Fujiwara K, Suematsu H, Aoki M, Sato M, Nobuko M. Size dependent toxicity of silica nanoparticles to Chlorella kessleri. J Environ Sci Health A 2008; 43: 1167-1173.

[115] van Hoecke K, De Schampelaere KAC, van der Meeren P, Loucas S, Janssen CR. Ecotoxicity of silica nanoparticles to the green alga Pseudokrirchneriella subcapita: importance of surface area. Environ Toxicol Chem 2009; 27: 1948-1957.

[116] Gur J, Wang ASS, Chen C, Jan K. Ultrafine titanium dioxide particles in the absence of photoactivation can induce oxidative damage to human bronchial epithelial cells. Toxicol 2005; 213: 66-73.

[117] Lanone S, Boczkowski J. Biomedical applications and potential health risks of nanomaterials: molecular mechanisms. Curr Mol Med 2006; 6: 651-663

[118] Nel A, Xia T, Madler L, Li N. Toxic potential of materials at the nanolevel. Science 2006; 311: 622-627.

[119] Jin C, Zhu B, Wang X, Lu Q. Cytotoxicity of titanium dioxide nanoparticles in mouse fibroblast cells. Chem Res Toxicol 2008; 21: 1871-1877.

[120] Kang JL, Moon C, Lee HS, *et al.* Comparison of biological activity between ultrafine titanium dioxide particles in RAW 264.7 cells associated with oxidative stress. J Toxicol Environ Health A 2008; 71: 478-485.

[121] Park E, Yi J, Chung K, Ryu D, Choi J, Park K. Oxidative stress and apoptosis induced by titanium dioxide nanoparticles in cultured BEAS-2B cells. Toxicol Lett 2008; 180: 222-229.

[122] Reeves JF, Davies SJ, Dodd NJF, Jha AN. Hydroxyl radicals are associated with titanium dioxide nanoparticle-induced cytotoxicity and oxidative damage in fish cells. Mutat Res 2008; 640: 113-122.

[123] Park E and Park K. Oxidative stress, pro-inflammatory responses induced by silica nanoparticles *in vivo* and *in vitro*. Toxicol Lett 2009; 184: 18-25.

[124] Liu S, Xu L, Zhang T, Ren G, Yang Z. Oxidative stress and apoptosis induced by nanosized titanium dioxide in PC 12 cells. Toxicol 2010: 267: 172-177.

[125] Warheit DB, Webb TR, Colvin VL, Reed KJ, Sayes CM. Pulmonary bioassay studies with nanoscale and fine- quartz particles in rats. Toxicity is not dependent on particle size but on surface characteristics. Toxicol Sci 2007; 95: 270-280.

[126] Lison D, Thomassen LCJ, Rabolli V, *et al.* Nominal and effective dosimetry of silica nanoparticles in cytotoxicity assays. Toxicol Sci 2008; 104: 155-162.

[127] Horie M, Nishio K, Fujita K, *et al.* Protein adsorption of ultrafine metaloxide and its influence on cytotoxicity towards cultured cells. Chem Res Toxicol 2009; 22: 543-553.

[128] Myllynen F. Damaging DNA from a distance. Nature Nanotchnol 2009; 4: 795-796

[129] Lu N, Zhu Z, Zhao X, Tao R, Yang S, Gao Z. Nano titanium dioxide photocatalytic protein tyrosine nitration: a potential hazard of TiO₂ on skin. Biochem Biophys Res Commun 2008; 370: 675-680.

[130] Helland A, Kastenholz H, Siegrist M. Precaution in practice. Perceptions, procedures and performance in the nanotech industry. J Ind Ecol 2008; 12: 449-458.

[131] Ferrari A. Developments in the debate on nanoethics: traditional approaches and the need for new kinds of analysis. Nanoethics 2010; 4: 27-52.

[132] Grieger KD, Baun A, Owen R. Redefining risk research priorities for nanomaterials. J Nanopart Res 2010;12: 383-392.

[133] van Broekhuizen P, Reijnders L. Trade unions and NGOs positioning in the nanotechnologies' debate in the EU - some experiences of the Nanocap project. Risk Analysis 2011; DOI: 10.111/j.1539-6924.2011, 01615.x.

[134] He X, Liu F, Wang K, Ge J, Qin D, Gong P, Yan W. Bioeffects of different functionalized silica nanoparticles on HeCaT cell line. Chin Sci Bull 2006; 51: 1939-1946.

[135] Fadeel B, Garcia-Bennet AE. Better safe than sorry: understanding the toxicological properties of inorganic nanoparticles manufactured for biomedical applications. Adv Drug Deliv Rev 2010; 62: 362-374.

[136] Ciftcioglu N, Aho KM, McKay DS, Kajander EO. Are apatite nanoparticles safe? Lancet 2007; 369: 2078.

[137] Yuan Y, Liu C, Qian J, Wang J, Zhang Y. Size-mediated cytotoxicity and apoptosis of hydroxyapatite nanoparticles in human hepatoma HepG2 cells. Biomater 2010; 31: 730-740.

[138] Carlotti ME, Ugazio E, Sapino S, Fenoglio I, Greco G, Fubini B. Role of particle coating in controlling skin damage photoinduced by titania nanoparticles. Free Radic Res 2009; 43: 312-322.

[139] Park J, Kwak BK, Bae E, *et al.* Characterization of exposure to silver nanoparticles in a manufacturing facility. J Nanopart Res 2009; 11: 1705-1712.

[140] Tsai SC, Ada E, Isaacs JA, Ellenbecker MJ. Airborne nanoparticle exposures associated with the manual handling of nanoaluminia and nanosilver in fume hoods. J Nanopart Res 2009; 11: 147-161.

[141] Johnson DR, Methner MM, Kennedy AJ, Steevens JA. Potential for occupational exposure to engineered carbon-based nanomaterials in environmental laboratory studies. Environ Health Persp 2010; 118: 49-54.

[142] Shaffer RR, Rengasamy S. Respiratory protection against airborne nanoparticles: a review. J Nanopart Res 2009; 11: 1661-1672.

CHAPTER 6

Molecular Methods for Nanotoxicology

Lisa Bregoli*, Stefano Pozzi-Mucelli and Laura Manodori

Veneto Nanotech, via San Crispino 106, Padua, Italy

Abstract: This chapter presents the most frequently used methods to study the effect of nanomaterials and nanoparticles on biological systems. The aim of the chapter is not to give a detailed technical description of the protocols, but to present the available techniques that have been adopted for the analysis of potential toxic effect of nanomaterials, and engineered nanoparticles in particular. As nanotoxicology is an extremely new branch of bio-toxicological sciences, the definition of its methods is still in the process of development. In this chapter we describe the main challenges of this process, where the adaptation of classical cytotoxicity and molecular methods has to take into account the unique properties of nanomaterials.

Keywords: Nanotoxicology, Nanoparticle toxicity, *In vitro* cytotoxicity, Sub-lethal toxicity assay, Nanoparticle characterization, Physico-chemical Characterization, Protein binding, Oxidative stress, Nanoparticle cellular localization, *In vitro* cancerogenicity, *In vitro* genotoxicity

INTRODUCTION

Current and future applications of nanotechnology are expected to hold great health, societal and environmental benefits. The potential hazards of manufactured nanomaterials have been debated in recent years, especially following a number of studies, which indicated that some nanomaterials can cause adverse effects on *in vitro* biological systems and laboratory animals [1-3]. Data on nanoparticles, such as increasing production volumes and commercialization, ability to cross biological barriers, and increased biological activities when compared to bulk counterparts, have caused worries about their potential impacts on the health and safety of both humans and the environment. Perhaps more than any preceding technology, nanotechnology has been characterized by discussions about potential risks since its early development, with the result that most economies investing in nanotechnology also implement the discussions with questions concerning potential risks and how to manage them. Government-led agencies have prompted initiatives for a harmonized control over the risks posed by nanotechnology, such as the United States' National Nanotechnology Initiative. In the UK, the Royal Society and Royal Academy of Engineering (RS & RAE) (2004) galvanized the development of cross-agency groups to address uncertainties regarding the risks of nanomaterials.

Risk assessment methodologies for nanomaterials are still a matter of debate. The international scientific community is currently in the process of discussing, evaluating and refining nanotechnology-specific risk assessment strategies, with the aim to be able to perform complete scientifically valid quantitative risk assessments of nanomaterials in the nearest possible future. Once such risk assessments strategies and methods will be widely accepted, they will lead to informed risk management decisions aimed at protecting human health and the environment while reaping the benefits of nanotechnology for society.

When risk assessment of nanomaterials is discussed, it is often in the context of previous experience with chemical risk assessment, consisting of four parts: hazard identification, dose-response assessment, exposure assessment, and risk characterization. In Europe, legislation for controlling the production, use and release of chemical substances is based on risk assessment, as described in detail in the "Technical Guidance Document" (TGD) (European Commission JRC 2003), which aims to help competent authorities

*Address correspondence to Lisa Bregoli: Veneto Nanotech, Via San Crispino, 106, I-35129 Padova, Italy; Tel: 0425 377 511; Fax: 0425 377 555; E-mail: lisa.bregoli@ecsin.eu

Haseeb Ahmad Khan and Ibrahim Abdulwahid Arif (Eds)

to carry out risk assessments. It includes extensive technical details for conducting hazard identification, dose (concentration) – response (effect) assessment, exposure assessment and risk characterization in relation to human health and the environment.

Hazard identification is defined as the "*identification of the adverse effects which a substance has an inherent capacity to cause*" (European Commission JRC 2003). Until recently the potential negative effects of nanomaterials on human health and the environment were rather speculative and unsubstantiated. This has changed within the past few years and a number of laboratory studies have indicated that exposure to some nanoparticles, such as carbon nanotubes, fullerenes and metal nanoparticles, can lead to adverse effects in the lungs and the brain of test animals [1-5].

According to the TGD, a dose-response assessment involves "...an estimation of the relationship between dose, or level of exposure to a substance, and the incidence and severity of an effect" (European Commission JRC 2003). Several of the studies mentioned above have reported such a relationship. This goes for, especially, *in vitro* studies on among other C60 fullerenes, single- and multiwalled carbon nanotubes, and various forms of nanometals. Classically, dose refers to 'dose by mass', however, based on the experiences gained in biological tests of nanoparticles, it has been suggested that biological activity of nanoparticles might not be mass-dependent, but dependent on physical and chemical properties not routinely considered in toxicity studies [6]. Which properties determine or influence the inherent hazards of nanoparticles is still an open question, partly due to the general lack of characterization of the tested nanoparticles. According to some investigators, the surface area of the nanoparticles is a better descriptor of the toxicity of low-soluble, low toxicity particles [7-10], whereas others found that the particle number worked best as dose metrics [11], or that toxicity was related to the number of functional groups in the surface of nanoparticles [12]. Physical and chemical properties such as particle size, size distribution, number concentration, agglomeration state, shape, crystal structure, chemical composition, surface area, surface chemistry, surface charge, porosity, and method of synthesis are in the literature proposed as properties that need to be considered. However, many of the proposed physical and chemical properties are overlapping, or are only applicable to nanoparticles and not to nanomaterials in general.

Exposure is a key element in risk assessment of nanomaterials, since it is the condition for the potential toxicological and ecotoxicological effects to take place. If there is no exposure – there is no risk. According to the Technical Guidance Document exposure assessment involves "...*an estimation of the concentrations/doses to which human populations (i.e., workers, consumers and man exposed indirectly via the environment) or environmental compartments (aquatic environment, terrestrial environment and air) are or may be exposed.*" (European Commission JRC 2003).

Completing a full exposure assessment requires extensive knowledge about among others manufacturing conditions, level of production, industrial applications and uses, consumer products and behaviour, and environmental fate and distribution. Such detailed information is not available and so far no full exposure assessment has been published for nanomaterials. This may partly be due to difficulties in monitoring nanomaterial exposure in the workplace and the environment, and partly due to the fact that the biological and environmental faith of nanomaterials are still largely unexplored [13].

The primary route of exposure in occupational settings is assumed to be through inhalation and/or dermal contact after the manufacturing process of a nanomaterial, for instance when a reaction chamber is opened, a product is dried, or during the handling of products after their manufacture. Exposure is less likely during the manufacturing process itself since most nano-manufacturing processes are performed in a closed reaction chamber. However, unexpected system failure such as rupture of a seal may happen. Occupational exposure to ultrafine particles has a long history but for the moment, it is unclear to what extent analogies can be drawn to engineered nanoparticles. Whereas the fraction of the total ultrafine particle number concentrations generally decreases, fine particle number concentrations increases with time and distance from the point of emission [14]. The information and data publicly available about current levels of worker exposure to nanomaterials is very limited. This includes valuable information such as what kinds of nanomaterials workers are exposed to, where and how, the concentrations by dose or by particles number they are exposed to and what kinds of protective measures are used or are available.

Environmental exposure of nanomaterials seems inevitable with the increasing production volumes and the increasing number of commercially available products containing nanomaterials. Environmental routes of exposure are multiple and can stem from:

- operations related to the production of nanomaterials such as cleaning of production chambers;

- spills from production, transport, and disposal of nanomaterials or products;

- the use and disposal of products containing nanoparticles including incomplete waste incineration and landfills;

- wastewater overflow and ineffective sewage treatment plants (STP) unable to hold nanoparticles back or degrade them;

- degradation of products containing nanomaterials.

The total load to the environment from current uses of nanomaterials is unclear and analytical methods to detect and quantify environmental concentrations of nanoparticles have yet to become available. However various estimates have been made both for individual products, nanomaterials and applications as well as product types.

PHYSICOCHEMICAL CHARACTERIZATION

Every nanotoxicity assessment ideally begins with the physico-chemical characterization of the tested nanomaterial, which includes a quantification of possible impurities or contaminants, because they can create artefacts and alter the tests results. Knowledge of the properties of the tested nanomaterial is necessary to be able to compare different studies, and to understand the relationship between the physical and chemical characteristics and potential adverse effects.

A general need for harmonization of the methodologies used for the characterization is strongly needed, and it is actively being tackled by the scientific community. Also discussions about which nanoparticles physico-chemical properties are important in the risk assessment of nanomaterials are still ongoing, although a consensus is emerging. According to a 2009 opinion of the European Scientific Committee on Emerging and Newly Identified Health Risks (SCENIHR) [15], in line with OECD and ISO, there is a list of chemical and physical properties to be included in a nanotoxicology study (see below).

Clearly, not all the listed properties can be determined in every case. Instead a combination should be chosen, that clearly and uniquely defines the nanomaterial. One important consideration to be made is that the properties of nanomaterial are largely dependent on the surrounding material/medium, which can also influence changes and variations over the course of time. This has to be taken into account during the definition of a nanotoxicology study, and the choice of the right physico-chemical characteristics to be determined in relation to the specific conditions and methods is of paramount importance.

The chemical properties to be included in a nanotoxicology study are:

- Structural formula/molecular structure

- Composition of nanomaterial (including degree of purity, known impurities or additives)

- Phase identity

- Surface chemistry (composition, charge, tension, reactive stress, physical structure, photocatalytic properties, zeta potential)

- Hydrophilicity/lipophilicity

Some of the available analysis techniques to determine the chemical composition and purity of nanomaterials and nanoparticles are listed, while we refer to different sources for a more in depth description of the techniques [16]:

- inductively coupled plasma-mass spectroscopy (ICP-MS), which is able to detect even trace metal impurities, but refers to the bulk chemical composition,

- Time-of-flight secondary ion mass spectrometry (ToF-SIMS),

- X-ray fluorescence (XRF),

- energy dispersive X-ray analysis (EDX),

- surface-enhanced Raman spectroscopy (SERS).

The main physical properties of interest with respect to nanoparticle safety are:

- Size, shape, specific surface area, aspect ratio

- Agglomeration/aggregation state

- Size distribution

- Surface morphology/topography

- Structure, including crystallinity and defect structure

- Solubility.

Nanoparticle sizing and aggregation are critical aspects of pre-characterization for nanotoxicity studies. By definition, nanomaterials possess at least one dimension in the range of 1-100 nm but, due to their intrinsic high dispersion and elevated surface energy, nanoparticle aggregation is common in biological fluids. Since aggregation can cause significant changes in surface area properties ad interaction with biological and cellular components, a measure of nanoparticles sizing is not complete without a quantification of aggregation in the relevant media.

Microscopy techniques allow to evaluate aggregation, shape, size and, when coupled with spectroscopic probes, chemical composition of nanoparticles. Depending on the technique, resolution can reach sub-nanometer range, but the statistical significance of the measurement may be invalidated by the number of particles observed. The most commonly used methods are transmission (TEM) (Fig. **1**), scanning electron microscopy (SEM), and atomic force microscopy (AFM), which generally need samples in the dry state.

Among spectroscopic techniques, UV-Vis and infrared are particularly suitable for quantum dots and organic-based nanoparticles, such as fullerenes C60 and carbon nanotubes. X-ray spectroscopy, which comprise a wide range of analytical tools, can reveal information about the elemental composition or the crystallographic structure. Beside these, dynamic light scattering (DLS) is a rapid, non-destructive and simple analysis to evaluate the size distribution of particles dispersed in liquid samples, and its range of measurements is very wide (from few nm to few micron). Other real-time techniques are available for fluorescent nanoparticles, such as fluorescence polarization and fluorescence correlation spectroscopy. The total surface area of solid samples can be evaluated by the Brunauer Emmett Teller (BET).

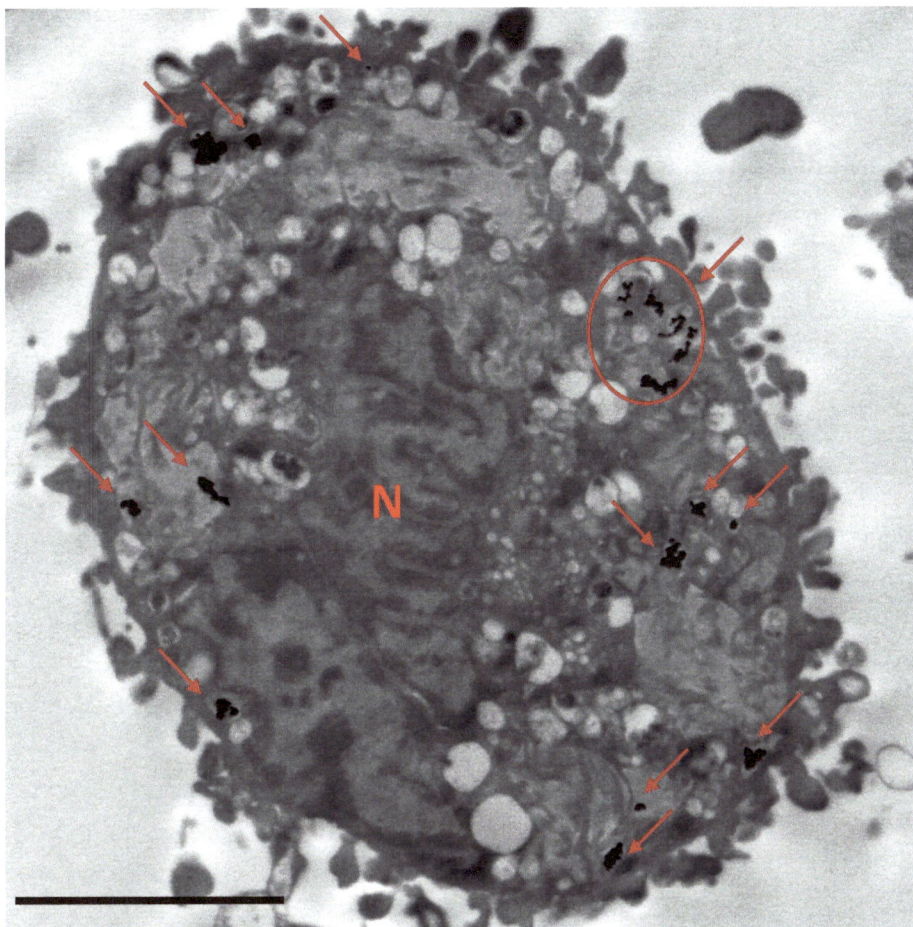

Figure 1: Transmission electron microscopy (TEM) image of a polymorphonuclear cell cultivated *in vitro* from blood progenitor cells. The bilobated nucleus (N) is visible in the center of the cell. Metal oxide nanoparticles that have been uptaken by the cell (pointed by red arrows) are present in vesicles. Size bar: 5 micron.

SURFACE CONTAMINATION

The importance of surface contamination detection and remediation has been neglected in several *in vitro* studies, thus making it impossible to determine whether the difference among observed responses of biological systems is due to the tested nanomaterials, the different cellular system, or a contamination of the nanoparticle stock solution. Surface contamination of materials is driven by thermodynamic laws, and it happens through adsorbtion of airborne or waterborne contaminants. Also, surface contamination can be due to a chemical process, such as surface oxidation, corrosion, charging or electron transfer. Given the high surface-volume ratio, which is typical of nanomaterials, surface contaminants potentially represent a considerable fraction, so that nanomaterial surface contamination can unintentionally introduce high levels of additional substances in a test system. If not detected, these forms of contamination can represent an unpredictable and confusing variable, since they can alter the biological response, and make it impossible to determine whether the observed effect is caused by the nanoparticle itself, the contaminants, or a synergy or anergy between the two. This in fact dramatically reduces the meaning and usefulness of a nanotoxicology test.

Surface contaminants can come from different sources, probably the most common being bacterial. Bacterial endotoxin, such as lipopolysaccharide (LPS), is a ubiquitous, heavily glycosylated, phosphorylated lipid which can virtually attach to any surface, since it can bind to both hydrophilic and positively charged surfaces. The effects of LPS on biological systems are relevant, as it is a pyrogen and induces a cellular response that can alter the production of inflammatory cytochines. For example, orthopedic wear particles have been shown to bind LPS and induce local inflammation, while the binding of this endotoxin to TiO_2 reduces the *in vitro* inflammatory response to LPS. It is therefore essential to determine the presence of such a contaminant on nanomaterials, lab glassware, lab supply water and cell culture materials, and be able to remediate before any experiment. The most widely used test assay for LPS contamination is the Limulus Amebocyte Lysate (LAL) assay. It is based on a colorimetric variation of the lysate, which is proportional to the amount of LPS activity [17]. One limitation is that it only detects endotoxin which is soluble or readily displaceable from the surface, but variations of this test have been developed, to correlate the amount of free to surface bound endotoxin [18].

Other possible surface contaminants are polyaromatic hydrocarbons (PAHs), which are ubiquitously present in air, and are by-products of production processes, for example in the production of carbon-based nanomaterials. Also volatile hydrocarbons, silicones, alkylated phthalates, and catalyst species or unreacted synthesis chemicals can be generated from nanomaterials production processes. These are all contaminants that can potentially alter the cellular response when added to the system, making it impossible to determine whether the observed effect is due to the physico-chemical characteristics of the nanomaterial *per se*.

Surface analysis techniques can be used to detect such surface contaminations. A list of such available techniques follows:

- Time-of-flight secondary ion mass spectrometry (ToF-SIMS),

- X-ray photoelectron spectroscopy (XPS),

- X-ray luorescence (XRF),

- energy dispersive X-ray analysis (EDX),

- surface-enhanced Raman spectroscopy (SERS).

Decontamination from this type of chemical contaminants is not easily performed by rinsing in water. The removal of catalysts from carbon nanotubes, such as Fe, Co and Ni, is usually done by stirring in acid for two days. Amorphous carbon, which is also a contaminant in the production of carbon nanotubes, is performed with thermical treatment or incubating in H_2O_2 for 20 to70 hours.

PARTICLE BEHAVIOUR IN BIOLOGICAL FLUIDS

Particle dispersion in biological fluids is a key issue when investigating the behavior and the interactions of the particles when they come in contact with biological systems. Despite the remarkable speed of development of nanoscience, relatively little is known about the interaction of nanoscale objects with biological systems, and this is now a serious bottleneck in the whole nanomedicine and nanotoxicology enterprise. For example, gene transfection and other forms of intracellular delivery depend on these issues, and rational approaches to that field have been limited by poor understanding of the nature of the surface of the transfection vector and how this affects its efficiency. When nanoparticles enter a biological fluid, they become coated with proteins that may transmit biological effects [19] due to altered protein conformation, exposure of novel epitopes, and perturbed function (due to structural effects or local high concentration).

A deep understanding of the biological effects of nanoparticles requires knowledge of the binding properties of proteins (and other molecules) that associate with the particles. However, the isolation and identification of particle-associated proteins, a fundamental prerequisite for nanotoxicology, is not a simple task. Furthermore, in terms of the biological response, the more abundantly associated proteins do not necessarily have the most profound effect. A less abundant protein with high affinity and specificity for a particular receptor may instead be a key player. It is thus essential to develop methods to identify both major and minor particle-associated proteins, and to study the competition between proteins to bind when the system is under kinetic or thermodynamic control. A central methodological problem is to separate free protein from protein bound to nanoparticles, ideally employing non-perturbing methods that do not disrupt the protein-particle complex or induce additional protein binding. The preferred method to-date has been centrifugation, identifying the major serum proteins albumin, IgG and fibrinogen as being associated with a wide range of particles of seemingly disparate molecular composition [20-23]. Due to its high abundance, albumin is almost always observed on particles and may be retrieved even if it has relatively low affinity. Other proteins observed with several particle types in these centrifugation assays are immunoglobulins, apolipoproteins and alpha-1-antitrypsin.

The understanding of protein–nanoparticle interactions and the biological consequences that arise from these interactions may be advanced if we find means to go beyond the mere identification of particle-associated proteins. Of highest relevance would be information on the binding affinities and stoichiometries for different combinations of proteins and nanoparticles, and ranking of the affinities of proteins that coexist in specific body fluids or cellular compartments. The rates by which different proteins bind to and dissociate from nanoparticles, *i.e.,* the time scales on which particle-associated proteins exchange with free proteins, are other critical parameters determining their interaction with receptors, and biological effects. A tightly associated protein that exchanges slowly may follow the particle if it enters from the extracellular fluid into an intracellular location, whereas a protein with fast exchange will be replaced by an intracellular protein in such processes. The biological outcome may also differ depending on the relative protein exchange rates between nanoparticles and cellular receptors. It is clear that, in understanding how particles will interact with cells, these issues, currently almost unstudied, are amongst the most fundamental.

The rates of association and dissociation are likely to vary quite considerably with protein and particle type. The lifetimes of typical protein–protein complexes can vary from microseconds to weeks, and protein–ligand complexes typically range in their lifetimes of microseconds to days. The association rate constants of some complexes approach the diffusion-controlled limit, whereas conformational changes upon binding may slow down the process by orders of magnitude. Although most kinetic studies of adsorbed proteins concern extended surfaces of larger particles, reported time scales of exchange of adsorbed proteins from silica, polymer and TiO_2 nanoparticles range from 100 s to many hours [24-28]. The affinities and/or exchange rates depend on molecular details and the stability of the protein toward unfolding [24, 29]. In this section, we will go in details about the techniques that are and may be applied for the characterization of the protein corona, both in the identification of the proteins bound to the surface of a particle in the biological fluid, and in the thorough investigation of the kinetics ruling these interactions as well.

Immunoprecipitation Assays

Immunoprecipitation is one of the methods of choice when investigating the interaction of a protein with other molecules, including nucleic acids, proteins, and other supramolecular complexes, when performed in native conditions. Operatively, the concept of the immunoprecipitation is simple: an antibody specific for the protein of interest is added to the protein lysate, the resulting sample is incubated for a sufficient time to allow the formation of the immunocomplex; then, taking advantage of bacterial proteins able to bind the Fc region of the antibodies, coupled to agarose or magnetic beads, it is possible to purify the immunocomplexes, and analyze them with conventional techniques [4]. Hence, if a particle is part of a protein complex, it is possible to precipitate this whole complex, verify its presence, and identify the proteins involved in the binding, after the disruption of the complex, by mean of immunological reactions, for known proteins, of mass spectrometry. However, several steps are to be considered, and optimized, in the process, to achieve consistent results and avoid experimental biases.

The sample preparation step needs to be performed in order to prevent complexes dissociation. This can be achieved firstly by operating in a cold room, or at least having care of leaving all the solutions and the samples always on ice, stabilizing the complexes; secondly, the appropriate buffer solutions need to be used, in particular avoiding the use of high concentration of detergents and reducing agents. Appropriate controls are crucial in the outcome of immunoprecipitation assays. Negative controls are essential for reliable immunoprecipitation assays. Their purpose is to verify that the protein(s) brought down in the experiment result from the specific interaction of the antibody with the antigen protein. The better control is to use isotype-matched immunoglobulin from the same species as the antibody being used for the IP, but directed to another, non relevant protein. Further, it is a good practice to include a negative control in which no antibody is added to the lysate, to rule out potential aspecific interactions between the beads and the lysate. This is particularly important when immunoprecipitating ENP-protein complexes, as it is difficult to predict how the particle will behave. Furthermore, a recent innovation in immunoprecipitation protocols is the use of magnetic beads, which allow fast protocols. Hence, the control step with no antibody is indeed of a great importance, as it can rule out not only aspecific molecular interactions, but magnetic interactions as well.

Gel Filtration

Gel filtration, also called size exclusion chromatography (SEC), is defined as "a separation technique in which separation mainly according to the hydrodynamic volume of the molecules or particles takes place in a porous non-adsorbing material with pores of approximately the same size as the effective dimensions in solution of the molecules to be separated". It is usually applied to large molecules or macromolecular complexes such as proteins and industrial polymers. Typically, when an aqueous solution is used to transport the sample through the column, the technique is known as gel filtration chromatography, versus the name gel permeation chromatography which is used when an organic solvent is used as a mobile phase. The main application of gel filtration chromatography is the fractionation of proteins and other water-soluble polymers, while gel permeation chromatography is used to analyze the molecular weight distribution of organic-soluble polymers. This is usually achieved with an apparatus called a column, which consists of a hollow tube tightly packed with extremely small porous polymer beads designed to have pores of different sizes. As the solution travels down the column some particles enter into the pores. Larger particles cannot enter into as many pores, hence the larger the particles, the faster the elution. The void volume includes any particles too large to enter the medium.

As described by Cedervall and coworkers [30], gel filtration can be used to study nanoparticle/protein complexes and their dissociation rates [31]. In details, the elution profile of a single protein in a protein–nanoparticle mixture depends on the dissociation rate, the chromatographic run time and flow rate, and other factors. If protein exchange from the particles is very slow, with a residence time several times longer than the separation time, one fraction of the protein would elute with the particles and one at the same position as for protein injected alone. If the exchange is very fast, the protein would elute at the same position as without particles. Intermediate dissociation rates produce divided or broadened peaks and the detailed elution pattern will be determined by the rates of protein–particle exchange.

This technique, in combination with other methodologies, may hence be helpful when studying a more intriguing aspect of protein-nanoparticle interaction, namely the equilibrium and kinetic binding properties of proteins (and other molecules) that associate with the particles. In particular, these methodologies include Isothermal Titration Calorimetry, and Surface Plasmon Resonance.

Isothermal Titration Calorimetry (ITC)

ITC [32] is a quantitative technique that can directly measure the binding affinity (Ka), enthalpy changes (ΔH), and binding stoichiometry (n) of the interaction between two or more molecules in solution. From these initial measurements Gibbs energy changes (ΔG), and entropy changes (ΔS), can be determined using the relationship:

$$\Delta G = -RTlnK = \Delta H - T\Delta S$$

(where R is the gas constant and T is the absolute temperature).

An isothermal titration calorimeter is composed of two identical cells made of a highly efficient thermal conducting material such as Hastelloy alloy or gold, surrounded by an adiabatic jacket. Sensitive circuits are used to detect temperature differences between the reference cell (filled with buffer or water) and the sample cell containing the macromolecule. Prior to addition of ligand, a constant power (<1 mW) is applied to the reference cell. During the experiment, ligand is titrated into the sample cell in precisely known aliquots, causing heat to be either taken up or evolved (depending on the nature of the reaction).

Measurements consist of the time-dependent input of power required to maintain equal temperatures between the sample and reference cells.

In an exothermic reaction, the temperature in the sample cell increases upon addition of ligand. This causes the feedback power to the sample cell to be decreased. Observations are plotted as the power in µcal/sec needed to maintain the reference and the sample cell at an identical temperature. This power is given as a function of time in seconds. As a result, the raw data for an experiment consists of a series of spikes of heat flow (power), with every spike corresponding to a ligand injection. These heat flow spikes/pulses are integrated with respect to time, giving the total heat effect per injection. The pattern of these heat effects as a function of the molar ratio [ligand]/[macromolecule] can then be analysed to give the thermodynamic parameters of the interaction under study. It should be noted that degassing samples is necessary in order to minimize data interference due to the presence of air bubbles within the cells. The presence of such bubbles will lead to abnormal data plots in the recorded results. The entire experiment takes place under computer control.

ITC can hence be applied to characterize the protein/nanoparticle thermodynamic binding parameters. In the previously cited work by Cedervall and coworkers, ITC has been applied to unravel differences in protein binding by two nanoparticles differing in hydrophilicity, by adding albumin to the nanoparticle solution in the sample cell of the calorimeter.

Surface Plasmon Resonance

Surface plasmons, also known as surface plasmon polaritons, are surface electromagnetic waves that propagate in a direction parallel to the metal/dielectric (or metal/vacuum) interface. Since the wave is on the boundary of the metal and the external medium (air or water for example), these oscillations are very sensitive to any change of this boundary, such as the adsorption of molecules to the metal surface. SPR is particularly promising in the field of nano-biosensors [33]; for nanoparticles, localized surface plasmon oscillations can give rise to the intense colors of suspensions or sols containing the nanoparticles. Nanoparticles or nanowires of noble metals exhibit strong absorption bands in the ultraviolet-visible light regime that are not present in the bulk metal. Shifts in this resonance due to changes in the local index of refraction upon adsorption to the nanoparticles can also be used to detect biopolymers such as DNA or proteins.

CHOICE OF *IN VITRO* SYSTEMS

Although there still are inconsistencies between *in vitro* models and *in vivo* observations, which decrease the predictability of *in vitro* systems, there is little rational or ethical justification to proceed directly from nanomaterial synthesis to *in vivo* testing. *In vitro* cell-based systems are actually considered central to toxicity, biomaterials and environmental material exposure testing. Several cell-based assays have been optimized for colloids and particulates (ISO 10993), and have been adopted to nanotoxicity testing. The adoption of "classical toxicity" methods to nanotoxicology has fuelled much debate in the last years, because nanomaterials physical properties, such as high surface reactivity, intrinsic photometric absorbance and fluorescence, interact with elements of the assays themselves, thus making the results often inconsistent and little informative. The chemical-physical nature and behaviour of nanomaterials is so complex, it has even subverted some of the most basic principles of "classical toxicology", such as that of "dose" and "concentration. It becomes then essential to be cautious about the choice of the assay outputs, which have to consider the possible interference of the nanomaterial with the read-out system, and more than one assay may be required when determining nanoparticle toxicity for risk assessment. Nevertheless, strong effort from the scientific international community is actually been made, to produce and optimize nanomaterial-specific *in vitro* cell-based testing procedures.

The choice of the most appropriate *in vitro* system usually mirrors the site of exposure of the tested nanoparticle, and both intentional and unintentional nanoparticle exposures have to be taken into consideration. In case of work exposure or air/pollution-borne nanoparticles, exposure is unintentional and the site of entry is mainly through inhalation, although dermal contact, as well as ingestion, are possible. For engineered nanoparticles that are added to commercial products in order to increase their performance, human exposure may be both intentional, for applications such as nanomedicine and dietary supplements, and unintentional as in the case of nanomaterial-doped textiles, households products and personal care products. The exposure for engineered nanoparticles may be by different routes, from intraperitoneal to intravenous (in the case of nanomedicine applications), dermal, inhalation, ingestion.

When nanoparticles enter the human body, they are able to cross the biological barriers and potentially reach every organ. Depending on their size, aggregation and chemical composition, nanoparticles have been shown to penetrate damaged skin, to cross the pulmonary, intestinal barriers and circulate in the blood. Metal nanoparticles, such as silver, have been shown to accumulate in the liver; it becomes than relevant for *in vitro* nanotoxicology testing to consider the cancerogenicity and toxicity potential in liver cells. Different nanoparticles have also been shown to reach the brain, by crossing the blood brain barrier, and have been seen in testis, although the demonstration that they have crossed the blood-testis barrier is still matter of research.

Depending on the route of exposure and potential organ targets, and the predicted concentration in the tissues, the most appropriate cell types and nanoparticle concentration range have to been chosen for toxicity testing. A wide range of cell types is now commercially available, both for cell lines and primary culture systems. Cell lines offer the advantage of lower technical and economical complexity for culture and higher expansion, although they originate either from cancerous tissue or immortalization processes, and the continuous growth causes genomic variations which make this culture system very different from the *in vivo* cells. The intrinsic value of primary cell cultures as predictors of *in vivo* cells response to nanoparticles is higher, as they are isolated from human tissues and have undergone considerably lower *in vitro* sophistications. Primary cells systems have the disadvantage of being more expensive, *in vitro* expansion is more limited and a higher expertise is required for their culture.

Cell Lines and Primary Cells

Macrophages are often used to test the potential of nanoparticles to induce the production of reactive oxygen species (ROS), oxidative stress, cell signalling such as calcium and NF-kB cariations, and cytokine production. Macrophages have also been shown to distinguish between pathogenic fibers, and can give an idea of the pro-inflammatory potential of nanoparticles. In the case of genotoxicity studies, on the contrary, macrophages are not the best choise, as more organ-specific cells have to be considered as target. Several other major cell types have been used, such as neural, hepatic, epithelial, endothelial and red blood cells, and several cancer cell lines [16, 34, 35].

Cell-nanomaterial effects that are usually examined are variations in viability, induction of necrosis or apoptosis, oxidative stress, increased production of reactive oxygen species (ROS), cell signalling, morphological alterations, membrane perturbation, mitochondria disfunction, production of cytochines, gene expression and other toxic reactions. Some of the cell lines and primary cells that have been used to assess the *in vitro* toxicity of nanomaterials (nanoparticles, quantum dots, gold nanoshells, fullerenes, carbon nanotubes) [16, 34] are listed in Table **1**.

Table 1: Some of the cell lines and human primary cells used for cytotoxicity studies for nanomaterials and nanoparticles [16, 34].

Cell lines and primary cells	Organ/tissue
Primary human umbilical vein endothelial cells [36] HUV-EC-C (umbilical vein) cell line	Endothelium (blood vessels)
HEK (human epidermal keratinocytes) [37-39] Human neonatal HEKs [40, 41] HDF (human dermal fibroblasts) [42-44] HaCaT (human epidermal keratinocytes) [45, 46] MSTO-211H (human mesothelioma cells) [47, 48] HSF42 (human skin fibroblasts)[49, 50]	Skin

B16F10 (mouse melanoma cells) [51] Melanoma cells [52]	
Primary human hepatocytes [53] HepG2 (Human liver carcinoma) [42, 54-59] BRL3A (Rat liver cells) [60]	Liver
Primary human lung epithelial cells [43] A549 (human alveolar epithelial cells from lung adenocarcinoma) [61-63] H596 (Human lung tumor) [64] H446 (Human lung tumor) [40, 64] Calu-1 (Human lung tumor) [40, 64] IMR-90 (Human embryonic lung fibroblasts) [50] MRC-9 (Human lung fibroblasts) [65]	Lung
HEK293 (human embryonic kidney) [66, 67] COS-7 (monkey kidney cells) [68, 69] COS-1 (monkey kidney cells) [70] Vero cells (monkey kidney epithelial cells) [53, 71, 72] BHK-21 (hamster kidney cell line) [73]	Kidney
NHA (Neuronal human astrocytes) [42] GS-9L (rat gliosarcoma) [74] RG-2 (rat glioma) [74] F-98 (rat glioma) [74] EC219 (Rat brain-derived endothelial cells) [75] N9 (mouse microglial cells) [75] N11 (mouse microglial cells) [75] SH-SY5Y (human neuroblastoma cells) [58] PC12 (rat pheochromocytoma) [76, 77]	Brain
CaCo2 (human colorectal adenocarcinoma) [78] MTX-E12 (mucus-secreting *in vitro* drug absorption model based on monolayers of goblet-cell like sub-clones of the human colon carcinoma cell line HT29) [78]	Intestine
Human T Lymphocytes [79] Primary human bone marrow hematopoietic progenitor/stem cells [80] Primary human bone marrow mesenchymal progenitor/stem cells [81] Human monocyte-derived macrophages [82-84] Macrophages and foam cells [85] Jurkat (Human T-cell leukemia) [79, 80] RAW264.7 (mouse macrophages) [86, 87] J6456 (mouse lymphoma) [88] THP-1 (human monocytic leukemia) [40, 63, 80, 89] Mono Mac (human monocytic leukemia) [63] K562 (Human leukemia cells) [80, 90] HL60 (human promyelocytic leukemia) [80]	Immune system/Blood/Bone marrow
NR8383 (Rat alveolar macrophages) [62] Guinea pig alveolar macrophages [91] RAW267.9 (mouse alveolar macrophages) [89]	Immune system/Lung
hTER-Bj1(human fibroblasts) [92] 3T6 (human fibroblasts) [93] 3T3 (mouse fibroblasts) [48, 56, 93, 94] Hela (human cervical cancer cells) [53, 56, 95-97] Human breast cancer cells [50] MDA-MB-231 (human breast cancer cell line) [98, 99] SK-BR-3 (human breast cancer cell line) [100-103] MCF-7 (human breast cancer cell line) [104] KB-FR (human head and neck KB cancer cell line) [88] Primary human osteoblast-like cells [105] MG63 (human osteosarcoma cell line) [105] Dendritic cells [106]	Others

Mouse primary peritoneal macrophages (from lavage)	
J774.1A (mouse peritoneal macrophage-like cells) [107]	
Sprague-Dawley rat peritoneal macrophages	
CHO (Chinese hamster ovary) [99]	

CYTOTOXICITY ASSAYS

Choice of Cytotoxicity Assays

As we mentioned in the previous paragraph, the choice of the cytotoxicity assay for nanotoxicity studies has to be made with caution, carefully considering the chemical-physical nature of the tested nanomaterial, which may interact with the read-out system. For all of the cytotoxicity assays that have colorimetric and fluorescent read-outs, internal positive and negative controls need to be optimized and included when adapting "standard" cytotoxicity assays to nanoparticle toxicity testing, which have to address the fact that nanoparticles may interfere. First of all, nanoparticles may generate an absorbance at the same wavelength as that used to quantify the colored product, leading to an overestimation of the cell viability. In some cases, this type of interference may be overcome by subtracting the background absorbance of the cells in presence of the particles, without the assay reagents. Also, the large surface area or other surface properties can result in high adsorptive capacity which causes the nanoparticles to effectively extract the colored product from the cell extract, leading to an underestimation of cell viability. In this case, the interference is more difficult to control for, and alternative assays should be considered. Additionally, nanoparticles can exhibit oxidative surface properties, and many of the colorimetric cytotoxicity assays are based on oxidative reactions. It is therefore essential to determine beforehand, whether the nanoparticle can oxidize the substrate in the absence of cell debris. In any way, even though the right choice if internal controls can make these assays more acceptable for nanotoxicity studies, the choice of multiple assay systems is often desirable.

Viability Assays

Several viability assays are available, which have been used for cytotoxicity assessment of nanomaterials [16, 34, 35]. The estimation of live versus dead cells number after exposure to a nanomaterial provides a first evaluation of a lethally toxic effect. Other parameters that are frequently used to quantify a nanomaterial-driven toxic effect are metabolic alterations, compared to non-treated cells. The most direct and fast way to test cell viability is through the use of dyes that allow the discernment of dead versus live cells, based on the uptake of the dye. Two such chemicals are Neutral Red and Trypan Blue. Neutral Red is a weak cationic dye that can cross the plasma membrane by diffusion; if the membrane is altered the uptake of the dye is decreased and it leaks out. Cytotoxicity can be quantified by taking spectrophotometric measurments of the neutral red under different conditions, versus control cells. Trypan Blue, on the contrary, is only permeable to cells with damaged membranes and binds to the DNA. With a simple light microscopy observation, the number of dead versus live cells can be quantified, as dead cells appear blue while live cells remain unstained.

Another assay that measures the number of damaged cells is the LIVE/DEAD viability assay. This includes two chemicals, calcein acetoxymethylester (calcein AM), an electrically neutral, esterified molecule, and ethidium homodimer. Calcein AM can easily enter cells by diffusion and, once within the cell, it is converted to the green fluorescent molecule calcein by esterases. Damaged or dead cells, instead, are stained by ethidium homodimer, a membrane-impermeable molecule, and fluoresce red. For nanomaterial-related toxicity, the lactate dehydrogenase (LDH) has also been used. In this assay, the LDH that has been released from the cytoplasm of damaged cells oxidizes lactate to pyruvate, which promotes the conversion of tetrazolium salt INT to formazan, a water-solublemolecule with absorbance at 490 nm. The amount of released LDH is proportional to the number of damaged cells.

Fluorescence activated cell sorting (FACS) is able to give viability information about a population of cells, as it can distinguish between healthy, apoptotic and necrotic cells. One fast and relatively easy technique that allows the distinction between necrosis and apoptosis is the double staining with propydium iodide and

Annexin V. The fluorescent dye propidium iodide (PI) stains the DNA of a dead cell, like the Trypan blue dye. This staining is used to discern between necrotic and apoptotic death, because an altered plasma membrane permeability is considered as a distinctive feature of necrosis. Annexin V binds to phosphatidyl serine on the surface of apoptotic cells. When PI is coupled to annexin V-FITC (fluorescein isothiocyanate) staining, in FACS experiments, it is possible to precisely measure the amount of necrotic versus apoptotic death. It also allows to have information about nanoparticle-driven alterations of the proportion of hypodiploid DNA cells in G0/G1 phases.

Other viability assays that indicate apoptotic DNA fragmentation and leakage from cells comprise the comet assay for the quantification of fragmented DNA with gel electrophoresis (see below), the caspase Glo3/7 that quantifies downstream effectors of the mitochondrial apoptotic pathway, the Hoechs staining that binds to double-stranded DNA that has leaked out from the nucleus and makes it detectable by fluorescence, the TUNEL measures the amount of fragmented DNA in the nucleus as a marker of apoptosis. As mentioned above, all the colorimetric and fluorescence assays need to be always carried out in parallel with a sufficient number of internal controls, to be able to quantify the degree of interference of the tested nanoparticles with the system readout.

Metabolic Activity Assays

Other colorimetric assays measure the metabolic activity, which often relates to cells vitality. One such assay is MTT, based on 3-(4,5-dimethylthiazol-2-yl)-2,5-diphenyl tetrazolium bromide. Mitochondrial dehydrogenases, which are only present in intact mitochondria, cleave the tetrazolium ring which turns to blue from its yellow original color therefore the reaction only occurs in living cells, which become blue. When applied to nanomaterial toxicology testing, the MTT assay needs to be optimized, based on the type of nanoparticles being analyzed. For example, cysteine-coated quantum dots are able to catalytically reduce MTT to formazan, without cellular metabolism. Other problems related to the use of MTT for nanotoxicity is that in the final stages of the assay, solubilization of cells and the formazan product is achieved with a solvent, such as dimethylsulfoxide (DMSO) of isopropanol. When testing nanoparticles, this generates a suspension containing cell debris, the dissolved formazan and particulates, which forms a background interference with the assay. In such a case, investigators [35] have found it useful to transfer the supernatant to a fresh 96-well plate, and therefore to read the absorbance of the supernatant devoid of particles and cell debris. Some variations of this assay exist, which are based on different substrates, such as the MTS assay, based on 3-(4,5-dimethylthiazol-2-yl)-5(3-carboxymethoxyphenyl)-2-(4-sulfophenyl)-2H-tetrazolium, and the WST assay. A different colorimetric assay is the Alamar blue, which is converted to a pink fluorescent dye by cell metabolic activity because it act as an electron acceptor for enzymes during oxygen consumption, such as NADP and FADH. The assessment of cellular adenosine triphosphate (ATP) content is also used, to determine the viability of cells. ATP content can be extracted with perchloric acid, followed by neutralization, and commercially available fluorescent kits allow the measurement of ATP content.

Sub-Lethal Toxicity Assays

Oxidative Stress

Not all nanomaterial-driven toxic effects lead to cell death, thus several assays have been designed to measure the "sub-lethal toxic effects" [16, 34, 35]. One such effect, which has been frequently reported in nanotoxicity studies, is the induction of oxidative stress, and it is often measured through the quantification of total glutathione (GSH) versus the oxidized form glutathione disulfide (GSSG). GSH is a major antioxidant compound that is oxidized to GSSG in the presence of reactive oxygen species (ROS); in order to sustain its role as antioxidant during oxidative stress, a high GSH/GSSG ratio is necessary. In response to an oxidative stress insult, the enzyme glutathione reductase becomes activated, which maintains GSH in its reduced form. One of the several glutathione assays detects levels of glutathione using the Ellman's reagent (5,5'-dithio-bis-2-nitrobenzoic acid, DTNB), which reacts with the sulphydryl group of GSH to produce a yellow-colored product (5-thio-2-nitrobenzoic acid, TNB). Glutathione reductase also recycles GSH from the GSH-TNB complex, producing more TNB. Absorbance at 405 nm or 412 nm is used to measure the amount of TNB, which is proportional to the GSH content in the sample. A different method to measure GSH is based on o-phthalaldehyde (OPT), which uses the same cellular extract required for the ATP assay and generates a fluorescent signal. An alternative method is to reduce the total glutathione using a reducing agent such as beta-mercaptoethanol, therefore allowing measurement of the ratio of GSH to toal glutathione.

Other markers of oxidative stress include measurements of lipid peroxidation of the plasma membrane, which can be detected with different assays, such as the TBA (thiobarbituric acid). This assay is based on MDA (malondialdehyde), a toxic byproduct of lipid peroxidation that reacts with TBA when exposed to heat and acidic pH, to form a fluorescent pink chromagen, which can be measured colorimetrically when excited at 532 nm. Alternatively, BODIPY-C_{11} is a fluorescent dye that inserts into lipid bilayers, allowing oxidized and unoxidized lipids to be quantified fluorimetrically and imaged by their respective green and red colors.

Gene- and protein expression modifications in response to an oxidative stress insult have also been put forward as a sensitive marker of oxidative stress induced by nanoparticles. In particular, one gene that is known to respond to oxidative stress is heme oxygenase-1 (HO-1). HO-1 is known to have antioxidative and anti-inflammatory properties, and its enhanced expression in the lung is widely recognized as a protective mechanism against oxidative tissue injury.

ROS Detection

Since the cell-based production of reactive oxygen species (ROS) is widely accepted as one of the major nanoparticle-specific effects, several assays have been developed and are frequently used in nanocytotoxicity experiments, which directly measure the amount of produced ROS [16, 34, 35]. Care must be exercised for ROS production assays with nanoparticles, such as carbon back and TiO_2 nanoparticles, that also generate ROS species in cell-free systems.

Many methods are available, with different sensitivities and specificities, as well as the ability to detect intracellular versus extracellular species. The assay based on fluorescent dye 2,7-dichlorofluorescin (DCFH) has been developed for nanoparticle testing by Foucaud and co-workers [108]. In this assay, the dye is obtained as a diacetate precursor and cleaved by high pH to make the non-fluorescent product DCFH. The presence of ROS converts DCFH to a fluorescent product which can be measured by fluorymetry. Nanoparticles can produce a background interference in the absence of the dyes, which needs to be assessed and deducted from the experimental reading. As ROS production tends to change over time, this reading has to be taken a different time points and the choice of the time point at which the reaction has not gone to completion as to be made, in order to compare different nanoparticles.

Electron Paramagnetic Resonance (EPR) has been used to detect ROS. This technique has the advantage of being an effective radical detection method in either the presence or absence of cells. For EPR detection of radicals, an adduct-forming, the spin-trap agent DMPO (5,5-dimethyl-1-pyrroline N-oxide) for hydroxide or superoxide radicals, or a radical-consuming spin probe (4-hydroxy-2,2,6,6-tetramethylpiperidine-1-oxyl) are introduced in the cell culture with the nanoparticles, for a certain amount of time. After the incubation, the supernatant is collected and analysed on an EPR spectrometer.

Alternative assays for the nanoparticle-induced ROS production are based on the phagocytic burst by neutrophils and macrophages. When a macrophage encounters a nanoparticle, it responds with induction of phagocytosis of the nanoparticle and production of ROS, such as superoxide anions, in the attempt to "destroy" the potential pathogen. The cytochrome C assay measures superoxide anion production by cells, and has been used to measure the response of monocytes-macrophages to nanotubes. The reduced and oxidized forms of cytochrome C are measured at different specific wavelength, to obtain the extent of oxidation. An alternative method involves the quantification of extracellular H_2O_2 by spectrophotometric determination of horseradish peroxidase-catalized oxidation of a specific probe. Luminescent assays are also available, which measure the phagocytic burst using chemical enhancers, such as lucigenin and luminol, and are especially relevant for fibre-like nanoparticles that tend to induce frustrated phagocytosis.

INDUCTION OF CYTOKINES AND OTHER PROTEINS

Several *in vitro* cellular responses to nanomaterials have been observed, which are detectable by analyzing the expression pattern of proteins such as, cytokines, chemokines, stress-response related proteins or other cell-type specific proteins. Here again, the physico-chemical nature of the tested nanomaterial must be taken into consideration, because nanoparticles have been shown to bind proteins. It is then possible, during the sample preparation for protein analysis, that some proteins are adsorbed on the surface of nanoparticles and cleared off by subsequent washing or sample clearing procedures. This possibility has to be tested experimentally, for each nanoparticle and protein of interest. In addition, the above described interference problems need to be considered, especially when the system readout is colorimetric or fluorescent. Protein expression pattern changes are detectable by means of fluorescence, western blot, 2D gel electrophoresis and enzyme-linked immunosorbent assay (ELISA), while alterations in the gene expression can be analyzed with assays based on polymerase chain reaction (PCR), quantitative PCR, microarray.

Several nanoparticles induce pro-inflammatory cytochines production and release in the culture medium, which can be detected and quantified by ELISA, a surface capture immune-meditated sandwich assay. Examples of cytokines and chemokines associated with nanoparticle-induced proingflammatory signaling are tumor necrosis factor alpha (TNF-alpha), interleukin (IL)8, IL1alpha, IL1beta, IL6, granulocyte macrophage colony stimulating factor (GM-CSF). These immunomodulatory proteins are produced and released in the cell culture medium, which is collected an incubated on ELISA plates that allow the binding, detection and quantification of the specific protein.

Gel electrophoresis is used to separate proteins from a complex pool, for example from a cellular or subcellular fraction lysate, based on their molecular weight. Proteins can then be transferred onto a nitrocellulose or PVDF membrane, where they become available for detection with the use of antibodies (Western Blot). Two-dimensional (2-D) electrophoresis allows a greater separation of proteins, which are run on two subsequent gels, based on isoelectric focusing and molecular weight. Individual protein spots, and particular protein post-translational modifications, can then be visualized with gel staining procedures, and alterations of the pattern of protein expression or modification between treated and untreated samples can be determined. Single protein spots can be analyzed with mass spectrometry techniques, in order to precisely identify the protein.

Fluorescence activated cell sorting (FACS) can also be used to quantify receptor expression by cells, to reveal mechanism of cell toxicity or death, like the upregulation of Fas receptor. FACS has the advantage of being able to quantify the expression of different proteins in one multi-color experiement, although the possible interference of nanoparticles with the readout system has to be accounted for. FACS has also been developed for the detection of cytokines, like the multiplexed cytometry bead assay (CBA), a distrinct application of FACS equipment, with the use of libraries of various fluorescen bead-linked antibodies for the simultaneous detection of 30 or more cytochines. Microarray-printed immunocapture multiplex assays are also available, for the detection of up to 25 cytokines in one experiment (for example, the Quansys Q-Plex™). Here, up to 25 antibodies are printed on the surface of separate wells, in a 96-wells plate.

Quantitative polymerase chain reaction (qPCR) and gene expression microarray techniques are able to detect changes in gene expression due to nanomaterial exposure, through the analysis of mRNA expression patterns. Differences in genes expression between control and treated cells can be used as indicative of an alteration of the cellular function or metabolism, in response to a nanoparticle. Yet, even though these techniques are very informative and allow a high throughput screening of the effects of several nanometarials in several cell lines, we need to keep in mind that the expression of mRNA does not necessarily mirror the expression of the encoded protein, since post-transcriptional modifications and other regulatory mechanisms come in play [109].

IN VITRO TESTS FOR CANCEROGENICITY AND GENOTOXICITY

One very promising method to test the cancerogenic potential induced by organic and inorganic compounds, which has been used for nanotoxicology studies, is the Balb/3T3 assay. Balb/3T3 are fibroblasts from mouse BALB/c embryos, obtained by several cell culture passages. The clone A31-1-1 has been selected, and it is used for the cell transformation assay (CTA) that allows an evaluation of both cytotoxicity and morphologic transformations. The assay requires about 10 days for the assessment of cytotoxicity, and a subsequent 5 weeks for the morphologic transformation. The cytotoxicity endpoint is the colony forming efficiency (CFE), while the formation of type III foci is considered as the neoplastic potential endpoint. Only type III foci are considered as tumorigenic, as they are able to induce cancer in nude mice, with a frequency of 85% [110].

The Balb3T3 assay is not able to discriminate among genotoxic cancerogenic agents and non-genotoxic cancerogenic agents. For this reason, it is useful to couple it to an *in vitro* genotoxicity assay. Many tests are available to screen for gene or chromosomal mutations, aneuploidy, DNA strand breacks, DNA adducts and induction of DNA repair genes. Currently, however, only limited data are available for *in vitro* genotoxicity testing strategies for nanoparticles, and three are the most frequently used assays: the salmonella reverse mutation (Ames test), the micronucleus and the comet assays.

The Ames test is based on bacterial cells, and it allows to detect the potential of a nanomaterial of causing DNA breaks, upon physical contact of the two entities. It uses amino-acid requiring strains of *Salmonella typhimurium* and *Escherichia coli* to detect point mutations, which involve substitution, addition or deletion of one or a few DNA base pairs. The principle of this bacterial reverse mutation test is that it detects mutations which revert mutations present in the test strains and restore the functional capability of the bacteria to synthesize an essential amino acid. The revertant bacteria are detected by their ability to grow in the absence of the amino acid required by the parent test strain. This assay is described in the Organization for Economic Co-operation and Development (OECD) Guideline 471, and has been used for the genotixicity testing of several nanoparticles, such as TiO_2, fullerenes, carbon nanotubes. One important drawback of this assay, is that it utilises prokaryotic cells, which differ from mammalian cells in such factors as uptake, metabolism, chromosome structure and DNA repair processes. The test therefore does not provide direct information on the mutagenic and carcinogenic potency of a substance in mammals, but is commonly employed as an initial screen for genotoxic activity and, in particular, for point mutation-inducing activity.

The most widely used mammalian cell-based assays for the evaluation of clastogenic (chromosome breaking) effects of nanoparticles are micronucleus assay (Fig. **2**) and alkaline comet assay (Fig. **3**). These have been proven to be adequate for *in vitro* genotoxicity testing of nanoparticles and fibers, and have been used for nanomaterials such as TiO_2, carbon black, Co-Cr alloy nanoparticles, carbon nanotubes.

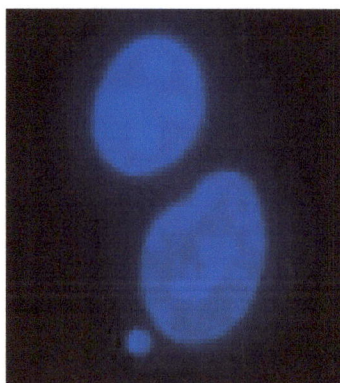

Figure 2: Micronucleus Assay: Fluorescent microscopy image of two nuclei with one micronucleus. The number of micronuclei formed is quantified after DNA-staining and is regarded as a measure for the potential of a substance to cause damage to the DNA. This photo was taken in the lab of Prof. Elke Dopp, University of Duisburg-Essen and herein published with her permission.

The purpose of the micronucleus assay is to detect those agents which modify chromosome structure and segregation in such a way as to lead to induction of micronuclei in interphase cells. It detects the formation of small membrane bound DNA fragments *i.e.,* micronuclei in the cytoplasm of interphase cells. These micronuclei may originate from acentric fragments (chromosome fragments lacking a centromere) or whole chromosomes which are unable to migrate with the rest of the chromosomes during the anaphase of cell division. After exposure to the nanoparticle, and addition of cytochalasin B for blocking cytokinesis cell cultures are grown for a period sufficient to allow chromosomal damage to lead to the formation of micronuclei in bi- or multi-nucleated interphase cells. Harvested and stained interphase cells are then analysed microscopically for the presence of micronuclei. Micronuclei are scored in those cells that complete nuclear division following exposure to the test item. Using fluorescent *in situ* hybridization (FISH) with probes targeted to the centromere region, it is possible to determine whether a specific micronucleus represents an acentric chromosome fragment (clastogenic effect) or whether it holds an entire chromosome (aneugenic effect). When using micronucleus assay for nanoparticles genotoxic assessment, the potential interaction of the nanoparticle with cytochalasin B must be considered and experimentally tested, and appropriate controls have to be run.

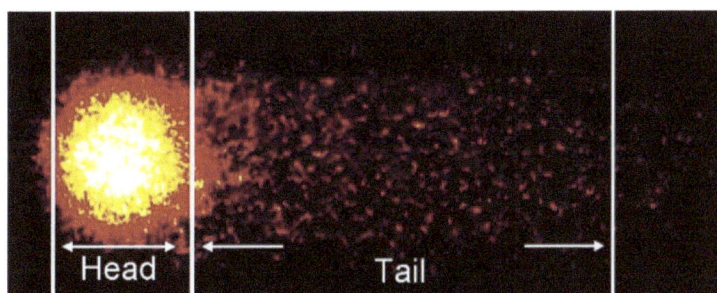

Figure 3: Comet Assay: DNA fragments migrate in an electric field out of the cell (head) into the tail. This picture is from the Department of Toxicology of the University of Würzburg, and is herein published with the Department's permission.

The comet assay is based on the microscopic detection of damaged DNA fragments of individual cells, which appear as comets upon cell lysis and subsequent DNA denaturation and electrophoresis. Various versions of this assay have been developed. The most common version allows the detection of single and double DNA strand breaks, DNA cross-links and alkali-labile sites. Other versions include the neutral comet assay for the quantification of DNA double strand breaks. Possible interference of nanoparticles with the comet assay readout has to be considered. For example, nanoparticles or aggregates can localize at or near comet appearances and affect their quantification due to fluorescence or ability to quench DNA-staining agents. It must also be considered that during the final processing steps of the assay, nanoparticles may come in contact with the nuclear DNA and thereby induce artificial damage.

Micronucleus and comet assays have specific and different advantages and disadvantages. The main strength of the micronucleus is that it can detect both chromosomal and genomic mutations, while its main limitation is that it can only be applied to dividing cells, in contrast to the comet assay. On the other hand, the comet assay is not able to detect fixed mutations, in contrast to the micronucleus assay. Appropriate positive and negative controls and optimization of the assays are necessary, in order to exclude interference due to the tested nanoparticle physico-chemical properties [10].

NANOPARTICLE CELLULAR LOCALIZATION

Microscopy techniques allow to track nanoparticles when they enter the cells, to identify their colocalization with biological molecules, and eventually to monitor parameters of cell activity after the intake of nanoparticles.

Nanoparticle Detection

Detecting, localizing and counting ultrasmall particles and nanoparticles in sub- and supra-cellular compartments are of considerable current interest in basic and applied research in biomedicine, bioscience and environmental science. For particles with sufficient contrast (*e.g.,* colloidal gold, ferritin, heavy metal-based nanoparticles), visualization requires the high resolutions achievable by transmission electron microscopy (TEM). TEM investigation can indeed provide several information on cellular localization, in particular showing with increasing degrees of detail the actual positions a nanoparticle may reach when they come in contact with the cells. L'Azou and coworkers [111] were able to estimate, for example, the differential intake and subcellular localization of carbon black and TiO_2 nanoparticles in renal cells, identifying size-dependent internalization processes, and furthermore being able to characterize the rate of aggregation of the particles and their subcellular localization, in this case in vacuolar structures.

For particles with electron densities not allowing their detection with electron microscopy, an alternative solution is the use of confocal or fluorescence microscopy, when fluorescently labeled particles were available. Anyway, these techniques have a resolution which is considerably higher than the conventional dimensions of a nanoparticle, being diffraction limits resolution to approximately 0.2 micrometer.

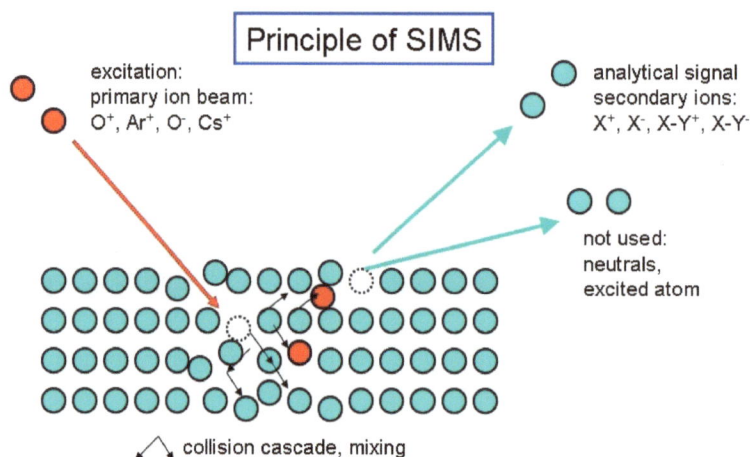

Figure 4: Principle of secondary ion mass spectrometry (SIMS). A high energy primary ion beam is directed at an area of the sample whose composition is to be determined. The bombardment of the sample surface is followed by mass spectrometry of the emitted secondary ions. Image is from Dr. Thomas Gemming, IFW Dresden, and herein published with his permission.

Imaging mass spectrometry can represent a valid solution to detect and visualize particles, independently from their electron density and eventual fluorescent labeling, but simply knowing their chemical composition. In particular, imaging implementation of secondary ion mass spectrometry (SIMS) presents the resolution necessary for the detection of nanoparticles in biological samples (Fig. **4**). In SIMS, a scanning ion beam is sputtered onto the surface to be analyzed, and the secondary ions ejected from the surface are collected and analyzed by a mass spectrometer, to determine the elemental, isotopic, or molecular composition of the surface. SIMS is the most sensitive surface analysis technique, being able to detect elements present in the parts per billion range. In imaging SIMS application, a thorough control of the primary ion beam positioning is needed, and the mass spectra resulting from the analysis are recorded keeping the information of their position of interest. This approach allows, when knowing a particular mass/charge ratio to investigate, to visualize the distribution of such m/z within the sample of interest. Hence, if a particular m/z can be foreseen for a given particle, it will be possible to investigate its distribution within a cell, a tissue, or another matrix. Furthermore, three dimensional recording can be performed, using the ion beam bombardment as a mean for depth profiling [112].

Nanoparticle Colocalization with Biological Molecules

The combination of microscopy techniques with immunological and genetic engineering techniques provides a mean to understand the colocalization and the interactions that particles may have with biological molecules. Immunohistochemistry, the use of antibodies specifically directed to cellular antigens, and appropriately labeled to be detected by the microscopes in use, is an approach used since decades in diagnostics and biological research. According to the microscopy technique used for the detection, the labeling of the antibody can occur by adding a strongly electron dense group, for electron microscopy applications, by conjugation with an enzyme (like peroxidase) which can catalyze a color producing reaction, for light microscopy, or even adding a fluorophore, to be detected by fluorescence and confocal microscopy. In nanotoxicology research, the use of immunohistochemistry in transmission electron microscopy can actually provide a strong information about the colocalization, due to the use of ultrafine sectioning of the biological samples: this approach limits indeed the displacement that the detected particle and the protein of interest may have on the z-axis, and the highest lateral resolution provides good information on the colocalization. When dealing with fluorescent particles, the use of confocal microscopy to detect antigen of interest or specific organelles (by mean of dyes targeted to these organelles), in successive z-sections of the sample, is another mean to prove the colocalization, with the advantage of a possible characterization of biological interactions in living cells.

Hence, Shi and coworkers [113] have been for example able to identify the lysosomal targeting of fluorescently labeled silica nanoparticle, and characterize their long term stability in living cells. A particular case is represented in such an approach by Quantum Dots (QDs). QDs are semiconductor nanocrystals, which emit light with an emission wavelength proportional to their size. They are of particular interest in biological imaging due to their brightness and stability, and since now have shown limited toxicity issues [58,114-116], which has been demonstrated to rely mainly on cadmium release and oxidative stress. Since QDs have this exceptional brightness, compared to conventional dyes, they have been proposed as ideal candidate as Förster resonance energy transfer (FRET) donors.

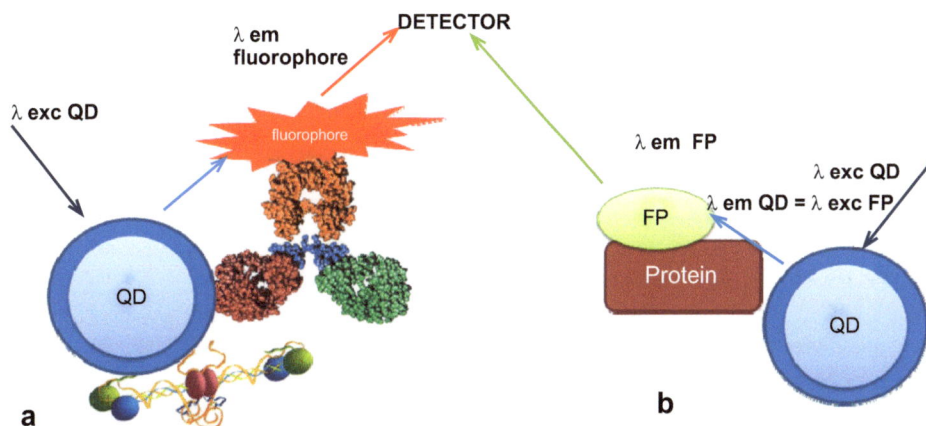

Figure 5: QDs in FRET applications, where the acceptin fluorophore can be covalently linked to an antibody directed to an interacting protein (a), or can be represented by a fluorescent protein coupled by genetic engineering techniques to the interacting protein (b).

In FRET, a donor chromophore, initially in its electronic excited state, may transfer energy to an acceptor chromophore (in proximity, typically less than 10 nm) through nonradiative dipole–dipole coupling. It is then possible to detect the light emitted from the accepting fluorophore, after exciting the donor fluorophore, if the two molecules are in a close contact, normally when molecules are interacting. This approach can be of an extreme interest when trying to understand the biological response of a cell to QDs, both by using antibodies directed to possible interacting proteins, but also when this protein are engineered with fluorescent proteins (Fig. **5**).

REFERENCES

[1]   Lam CW, James JT, McCluskey R, Hunter RL. Pulmonary toxicity of single-wall carbon nanotubes in mice 7 and 90 days after intratracheal instillation. Toxicol Sci 2004 Jan; 77(1): 126-34.
[2]   Oberdorster E. Manufactured nanomaterials (fullerenes, C60) induce oxidative stress in the brain of juvenile largemouth bass. Environ Health Perspect 2004 Jul; 112(10): 1058-62.

[3]   Poland CA, Duffin R, Kinloch I, *et al.* Carbon nanotubes introduced into the abdominal cavity of mice show asbestos-like pathogenicity in a pilot study. Nat Nanotechnol 2008 Jul; 3(7): 423-8.

[4]   Alber F, Dokudovskaya S, Veenhoff LM, *et al.* Determining the architectures of macromolecular assemblies. Nature 2007 Nov 29; 450(7170): 683-94.

[5]   Lewinski N, Colvin V, Drezek R. Cytotoxicity of nanoparticles. Small 2008 Jan; 4(1): 26-49.

[6]   Oberdorster G, Maynard A, Donaldson K, Castranova V, Fitzpatrick J, Ausman K, *et al.* Principles for characterizing the potential human health effects from exposure to nanomaterials: elements of a screening strategy. Part Fibre Toxicol 2005 Oct 6; 2: 8.

[7]   Oberdorster G. Significance of particle parameters in the evaluation of exposure-dose-response relationships of inhaled particles. Inhal Toxicol 1996; 8 Suppl: 73-89.

[8]   Oberdorster G, Oberdorster E, Oberdorster J. Concepts of nanoparticle dose metric and response metric. Environ Health Perspect 2007 Jun; 115(6): A290.

[9]   Stoeger T, Reinhard C, Takenaka S, *et al.* Instillation of six different ultrafine carbon particles indicates a surface area threshold dose for acute lung inflammation in mice. Environ Health Perspect 2006 Mar; 114(3): 328-33.

[10]   Stoeger T, Schmid O, Takenaka S, Schulz H. Inflammatory response to TiO2 and carbonaceous particles scales best with BET surface area. Environ Health Perspect 2007 Jun; 115(6): A290-A291.

[11]   Wittmaack K. In search of the most relevant parameter for quantifying lung inflammatory response to nanoparticle exposure: particle number, surface area, or what? Environ Health Perspect 2007 Feb; 115(2): 187-94.

[12]   Warheit DB. How meaningful are the results of nanotoxicity studies in the absence of adequate material characterization? Toxicol Sci 2008 Feb; 101(2): 183-5.

[13]   The Council of Canadian Academies. Small Is Different: A Science Perspective On The Regulatory Challenges of the Nanoscale. 2008.

[14]   Biswas P, Wu CY. Nanoparticles and the environment. J Air Waste Manag Assoc 2005 Jun; 55(6): 708-46.

[15]   SCENIHR (SCIENTIFIC COMMITTEE ON EMERGING AND NEWLY-IDENTIFIED HEALTH RISKS) EC. Risk Assessment of Products of Nanotechnologies. 2009.

[16]   Jones CF, Grainger DW. *In vitro* assessments of nanomaterial toxicity. Adv Drug Deliv Rev 2009 Jun 21; 61(6): 438-56.

[17]   Gorbet MB, Sefton MV. Endotoxin: the uninvited guest. Biomaterials 2005 Dec; 26(34): 6811-7.

[18]   Ratner BD, Bryant SJ. Biomaterials: where we have been and where we are going. Annu Rev Biomed Eng 2004; 6: 41-75.

[19]   Geys J, Nemmar A, Verbeken E, *et al.* Acute toxicity and prothrombotic effects of quantum dots: impact of surface charge. Environ Health Perspect 2008 Dec; 116(12): 1607-13.

[20]   Allemann E, Gravel P, Leroux JC, Balant L, Gurny R. Kinetics of blood component adsorption on poly(D,L-lactic acid) nanoparticles: evidence of complement C3 component involvement. J Biomed Mater Res 1997 Nov; 37(2): 229-34.

[21]   Goppert TM, Muller RH. Polysorbate-stabilized solid lipid nanoparticles as colloidal carriers for intravenous targeting of drugs to the brain: comparison of plasma protein adsorption patterns. J Drug Target 2005 Apr; 13(3): 179-87.

[22]   Kasche V, de BM, Lazo C, Gad M. Direct observation of intraparticle equilibration and the rate-limiting step in adsorption of proteins in chromatographic adsorbents with confocal laser scanning microscopy. J Chromatogr B Analyt Technol Biomed Life Sci 2003 Jun 25; 790(1-2): 115-29.

[23]   Labarre D, Vauthier C, Chauvierre C, Petri B, Muller R, Chehimi MM. Interactions of blood proteins with poly(isobutylcyanoacrylate) nanoparticles decorated with a polysaccharidic brush. Biomaterials 2005 Aug; 26(24): 5075-84.

[24]   Lundqvist M, Sethson I, Jonsson BH. Protein adsorption onto silica nanoparticles: conformational changes depend on the particles' curvature and the protein stability. Langmuir 2004 Nov 23; 20(24): 10639-47.

[25]   Lundqvist M, Sethson I, Jonsson BH. High-resolution 2D 1H-15N NMR characterization of persistent structural alterations of proteins induced by interactions with silica nanoparticles. Langmuir 2005 Jun 21; 21(13): 5974-9.

[26]   Renner L, Jorgensen B, Markowski M, Salchert K, Werner C, Pompe T. Control of fibronectin displacement on polymer substrates to influence endothelial cell behaviour. J Mater Sci Mater Med 2004 Apr; 15(4): 387-90.

[27]   Renner L, Pompe T, Salchert K, Werner C. Dynamic alterations of fibronectin layers on copolymer substrates with graded physicochemical characteristics. Langmuir 2004 Mar 30; 20(7): 2928-33.

[28]   Sousa SR, Moradas-Ferreira P, Saramago B, Melo LV, Barbosa MA. Human serum albumin adsorption on TiO2 from single protein solutions and from plasma. Langmuir 2004 Oct 26; 20(22): 9745-54.

[29] Lee WK, McGuire J, Bothwell MK. Competitive adsorption of bacteriophage T4 lysozyme stability variants at hydrophilic glass surfaces. J Colloid Interface Sci 2004 Jan 1; 269(1): 251-4.

[30] Cedervall T, Lynch I, Lindman S, *et al.* Understanding the nanoparticle-protein corona using methods to quantify exchange rates and affinities of proteins for nanoparticles. Proc Natl Acad Sci USA 2007 Feb 13; 104(7): 2050-5.

[31] Stevens FJ. Analysis of protein-protein interaction by simulation of small-zone size exclusion chromatography. Stochastic formulation of kinetic rate contributions to observed high-performance liquid chromatography elution characteristics. Biophys J 1989 Jun; 55(6): 1155-67.

[32] O'Brien R, Ladbury JE, Chowdry B.Z. Isothermal titration calorimetry of biomolecules. Oxford University Press; 2000.

[33] Hiep HM, Endo T, Saito M, *et al.* Label-free detection of melittin binding to a membrane using electrochemical-localized surface plasmon resonance. Anal Chem 2008 Mar 15; 80(6): 1859-64.

[34] Lewinski N, Colvin V, Drezek R. Cytotoxicity of nanoparticles. Small 2008 Jan; 4(1): 26-49.

[35] Stone V, Johnston H, Schins RP. Development of *in vitro* systems for nanotoxicology: methodological considerations. Crit Rev Toxicol 2009; 39(7): 613-26.

[36] Yamawaki H, Iwai N. Cytotoxicity of water-soluble fullerene in vascular endothelial cells. Am J Physiol Cell Physiol 2006 Jun; 290(6): C1495-C1502.

[37] Ryman-Rasmussen JP, Riviere JE, Monteiro-Riviere NA. Surface coatings determine cytotoxicity and irritation potential of quantum dot nanoparticles in epidermal keratinocytes. J Invest Dermatol 2007 Jan; 127(1): 143-53.

[38] Rouse JG, Yang J, Barron AR, Monteiro-Riviere NA. Fullerene-based amino acid nanoparticle interactions with human epidermal keratinocytes. Toxicol *In vitro* 2006 Dec; 20(8): 1313-20.

[39] Tian F, Cui D, Schwarz H, Estrada GG, Kobayashi H. Cytotoxicity of single-wall carbon nanotubes on human fibroblasts. Toxicol *In vitro* 2006 Oct; 20(7): 1202-12.

[40] Sato Y, Yokoyama A, Shibata K, *et al.* Influence of length on cytotoxicity of multi-walled carbon nanotubes against human acute monocytic leukemia cell line THP-1 *in vitro* and subcutaneous tissue of rats *in vivo*. Mol Biosyst 2005 Jul; 1(2): 176-82.

[41] Witzmann FA, Monteiro-Riviere NA. Multi-walled carbon nanotube exposure alters protein expression in human keratinocytes. Nanomedicine 2006 Sep; 2(3): 158-68.

[42] Sayes CM, Gobin AM, Ausman KD, Mendez J, West JL, Colvin VL. Nano-C60 cytotoxicity is due to lipid peroxidation. Biomaterials 2005 Dec; 26(36): 7587-95.

[43] Sayes CM, Wahi R, Kurian PA, *et al.* Correlating nanoscale titania structure with toxicity: a cytotoxicity and inflammatory response study with human dermal fibroblasts and human lung epithelial cells. Toxicol Sci 2006 Jul; 92(1): 174-85.

[44] Pernodet N, Fang X, Sun Y, *et al.* Adverse effects of citrate/gold nanoparticles on human dermal fibroblasts. Small 2006 Jun; 2(6): 766-73.

[45] Shvedova AA, Castranova V, Kisin ER, *et al.* Exposure to carbon nanotube material: assessment of nanotube cytotoxicity using human keratinocyte cells. J Toxicol Environ Health A 2003 Oct 24; 66(20): 1909-26.

[46] Manna SK, Sarkar S, Barr J, *et al.* Single-walled carbon nanotube induces oxidative stress and activates nuclear transcription factor-kappaB in human keratinocytes. Nano Lett 2005 Sep; 5(9): 1676-84.

[47] Wick P, Manser P, Limbach LK, *et al.* The degree and kind of agglomeration affect carbon nanotube cytotoxicity. Toxicol Lett 2007 Jan 30; 168(2): 121-31.

[48] Brunner TJ, Wick P, Manser P, *et al. In vitro* cytotoxicity of oxide nanoparticles: comparison to asbestos, silica, and the effect of particle solubility. Environ Sci Technol 2006 Jul 15; 40(14): 4374-81.

[49] Ding L, Stilwell J, Zhang T, *et al.* Molecular characterization of the cytotoxic mechanism of multiwall carbon nanotubes and nano-onions on human skin fibroblast. Nano Lett 2005 Dec; 5(12): 2448-64.

[50] Zhang T, Stilwell JL, Gerion D, *et al.* Cellular effect of high doses of silica-coated quantum dot profiled with high throughput gene expression analysis and high content cellomics measurements. Nano Lett 2006 Apr; 6(4): 800-8.

[51] Voura EB, Jaiswal JK, Mattoussi H, Simon SM. Tracking metastatic tumor cell extravasation with quantum dot nanocrystals and fluorescence emission-scanning microscopy. Nat Med 2004 Sep; 10(9): 993-8.

[52] Petri-Fink A, Chastellain M, Juillerat-Jeanneret L, Ferrari A, Hofmann H. Development of functionalized superparamagnetic iron oxide nanoparticles for interaction with human cancer cells. Biomaterials 2005 May; 26(15): 2685-94.

[53] Shiohara A, Hoshino A, Hanaki K, Suzuki K, Yamamoto K. On the cyto-toxicity caused by quantum dots. Microbiol Immunol 2004; 48(9): 669-75.

[54]   Hu Y, Xie J, Tong YW, Wang CH. Effect of PEG conformation and particle size on the cellular uptake efficiency of nanoparticles with the HepG2 cells. J Control Release 2007 Mar 12; 118(1): 7-17.

[55]   Tkachenko AG, Xie H, Coleman D, *et al.* Multifunctional gold nanoparticle-peptide complexes for nuclear targeting. J Am Chem Soc 2003 Apr 23; 125(16): 4700-1.

[56]   Tkachenko AG, Xie H, Liu Y, *et al.* Cellular trajectories of peptide-modified gold particle complexes: comparison of nuclear localization signals and peptide transduction domains. Bioconjug Chem 2004 May; 15(3): 482-90.

[57]   Liu Y, Chen W, Joly AG, *et al.* Comparison of water-soluble CdTe nanoparticles synthesized in air and in nitrogen. J Phys Chem B 2006 Aug 31; 110(34): 16992-7000.

[58]   Choi AO, Cho SJ, Desbarats J, Lovric J, Maysinger D. Quantum dot-induced cell death involves Fas upregulation and lipid peroxidation in human neuroblastoma cells. J Nanobiotechnology 2007; 5: 1.

[59]   Zhang Y, Chen W, Zhang J, Liu J, Chen G, Pope C. *In vitro* and *in vivo* toxicity of CdTe nanoparticles. J Nanosci Nanotechnol 2007 Feb; 7(2): 497-503.

[60]   Hussain SM, Hess KL, Gearhart JM, Geiss KT, Schlager JJ. *In vitro* toxicity of nanoparticles in BRL 3A rat liver cells. Toxicol *In vitro* 2005 Oct; 19(7): 975-83.

[61]   Worle-Knirsch JM, Pulskamp K, Krug HF. Oops they did it again! Carbon nanotubes hoax scientists in viability assays. Nano Lett 2006 Jun; 6(6): 1261-8.

[62]   Pulskamp K, Diabate S, Krug HF. Carbon nanotubes show no sign of acute toxicity but induce intracellular reactive oxygen species in dependence on contaminants. Toxicol Lett 2007 Jan 10; 168(1): 58-74.

[63]   Wottrich R, Diabate S, Krug HF. Biological effects of ultrafine model particles in human macrophages and epithelial cells in mono- and co-culture. Int J Hyg Environ Health 2004 Sep; 207(4): 353-61.

[64]   Magrez A, Kasas S, Salicio V, *et al.* Cellular toxicity of carbon-based nanomaterials. Nano Lett 2006 Jun; 6(6): 1121-5.

[65]   Limbach LK, Li Y, Grass RN, Brunner TJ, Hintermann MA, Muller M, *et al.* Oxide nanoparticle uptake in human lung fibroblasts: effects of particle size, agglomeration, and diffusion at low concentrations. Environ Sci Technol 2005 Dec 1; 39(23): 9370-6.

[66]   Cui D, Tian F, Ozkan CS, Wang M, Gao H. Effect of single wall carbon nanotubes on human HEK293 cells. Toxicol Lett 2005 Jan 15; 155(1): 73-85.

[67]   Salem AK, Searson PC, Leong KW. Multifunctional nanorods for gene delivery. Nat Mater 2003 Oct; 2(10): 668-71.

[68]   Thomas M, Klibanov AM. Conjugation to gold nanoparticles enhances polyethylenimine's transfer of plasmid DNA into mammalian cells. Proc Natl Acad Sci USA 2003 Aug 5; 100(16): 9138-43.

[69]   Cheng FY, Su CH, Yang YS, *et al.* Characterization of aqueous dispersions of Fe(3)O(4) nanoparticles and their biomedical applications. Biomaterials 2005 Mar; 26(7): 729-38.

[70]   Goodman CM, McCusker CD, Yilmaz T, Rotello VM. Toxicity of gold nanoparticles functionalized with cationic and anionic side chains. Bioconjug Chem 2004 Jul; 15(4): 897-900.

[71]   Su CH, Sheu HS, Lin CY, *et al.* Nanoshell magnetic resonance imaging contrast agents. J Am Chem Soc 2007 Feb 21; 129(7): 2139-46.

[72]   Hanaki K, Momo A, Oku T, *et al.* Semiconductor quantum dot/albumin complex is a long-life and highly photostable endosome marker. Biochem Biophys Res Commun 2003 Mar 14; 302(3): 496-501.

[73]   Mishra S, Webster P, Davis ME. PEGylation significantly affects cellular uptake and intracellular trafficking of non-viral gene delivery particles. Eur J Cell Biol 2004 Apr; 83(3): 97-111.

[74]   De Juan BS, Von BH, Gelperina SE, Kreuter J. Cytotoxicity of doxorubicin bound to poly(butyl cyanoacrylate) nanoparticles in rat glioma cell lines using different assays. J Drug Target 2006 Nov; 14(9): 614-22.

[75]   Cengelli F, Maysinger D, Tschudi-Monnet F, *et al.* Interaction of functionalized superparamagnetic iron oxide nanoparticles with brain structures. J Pharmacol Exp Ther 2006 Jul; 318(1): 108-16.

[76]   Pisanic TR, Blackwell JD, Shubayev VI, Finones RR, Jin S. Nanotoxicity of iron oxide nanoparticle internalization in growing neurons. Biomaterials 2007 Jun; 28(16): 2572-81.

[77]   Lovric J, Cho SJ, Winnik FM, Maysinger D. Unmodified cadmium telluride quantum dots induce reactive oxygen species formation leading to multiple organelle damage and cell death. Chem Biol 2005 Nov; 12(11): 1227-34.

[78]   Behrens I, Pena AI, Alonso MJ, Kissel T. Comparative uptake studies of bioadhesive and non-bioadhesive nanoparticles in human intestinal cell lines and rats: the effect of mucus on particle adsorption and transport. Pharm Res 2002 Aug; 19(8): 1185-93.

[79]   Bottini M, Bruckner S, Nika K, *et al.* Multi-walled carbon nanotubes induce T lymphocyte apoptosis. Toxicol Lett 2006 Jan 5; 160(2): 121-6.

[80] Bregoli L, Chiarini F, Gambarelli A, *et al.* Toxicity of antimony trioxide nanoparticles on human hematopoietic progenitor cells and comparison to cell lines. Toxicology 2009 Aug 3; 262(2): 121-9.

[81] Hsieh SC, Wang FF, Lin CS, Chen YJ, Hung SC, Wang YJ. The inhibition of osteogenesis with human bone marrow mesenchymal stem cells by CdSe/ZnS quantum dot labels. Biomaterials 2006 Mar; 27(8): 1656-64.

[82] Fiorito S, Serafino A, Andreola F, Togna A, Togna G. Toxicity and biocompatibility of carbon nanoparticles. J Nanosci Nanotechnol 2006 Mar; 6(3): 591-9.

[83] Porter AE, Muller K, Skepper J, Midgley P, Welland M. Uptake of C60 by human monocyte macrophages, its localization and implications for toxicity: studied by high resolution electron microscopy and electron tomography. Acta Biomater 2006 Jul; 2(4): 409-19.

[84] Muller K, Skepper JN, Posfai M, *et al.* Effect of ultrasmall superparamagnetic iron oxide nanoparticles (Ferumoxtran-10) on human monocyte-macrophages *in vitro*. Biomaterials 2007 Mar; 28(9): 1629-42.

[85] Chono S, Morimoto K. Uptake of dexamethasone incorporated into liposomes by macrophages and foam cells and its inhibitory effect on cellular cholesterol ester accumulation. J Pharm Pharmacol 2006 Sep; 58(9): 1219-25.

[86] Shukla R, Bansal V, Chaudhary M, Basu A, Bhonde RR, Sastry M. Biocompatibility of gold nanoparticles and their endocytotic fate inside the cellular compartment: a microscopic overview. Langmuir 2005 Nov 8; 21(23): 10644-54.

[87] Hu F, Neoh KG, Cen L, Kang ET. Cellular response to magnetic nanoparticles "PEGylated" *via* surface-initiated atom transfer radical polymerization. Biomacromolecules 2006 Mar; 7(3): 809-16.

[88] Schroeder JE, Shweky I, Shmeeda H, Banin U, Gabizon A. Folate-mediated tumor cell uptake of quantum dots entrapped in lipid nanoparticles. J Control Release 2007 Dec 4; 124(1-2): 28-34.

[89] Murr LE, Garza KM, Soto KF, *et al.* Cytotoxicity assessment of some carbon nanotubes and related carbon nanoparticle aggregates and the implications for anthropogenic carbon nanotube aggregates in the environment. Int J Environ Res Public Health 2005 Apr; 2(1): 31-42.

[90] Connor EE, Mwamuka J, Gole A, Murphy CJ, Wyatt MD. Gold nanoparticles are taken up by human cells but do not cause acute cytotoxicity. Small 2005 Mar; 1(3): 325-7.

[91] Jia G, Wang H, Yan L, *et al.* Cytotoxicity of carbon nanomaterials: single-wall nanotube, multi-wall nanotube, and fullerene. Environ Sci Technol 2005 Mar 1; 39(5): 1378-83.

[92] Gupta AK, Gupta M. Synthesis and surface engineering of iron oxide nanoparticles for biomedical applications. Biomaterials 2005 Jun; 26(18): 3995-4021.

[93] Pantarotto D, Briand JP, Prato M, Bianco A. Translocation of bioactive peptides across cell membranes by carbon nanotubes. Chem Commun (Camb) 2004 Jan 7; (1): 16-7.

[94] Chung YC, Chen IH, Chen CJ. The surface modification of silver nanoparticles by phosphoryl disulfides for improved biocompatibility and intracellular uptake. Biomaterials 2008 Apr; 29(12): 1807-16.

[95] Takahashi H, Niidome Y, Niidome T, Kaneko K, Kawasaki H, Yamada S. Modification of gold nanorods using phosphatidylcholine to reduce cytotoxicity. Langmuir 2006 Jan 3; 22(1): 2-5.

[96] Niidome T, Yamagata M, Okamoto Y, *et al.* PEG-modified gold nanorods with a stealth character for *in vivo* applications. J Control Release 2006 Sep 12; 114(3): 343-7.

[97] Duan H, Nie S. Cell-penetrating quantum dots based on multivalent and endosome-disrupting surface coatings. J Am Chem Soc 2007 Mar 21; 129(11): 3333-8.

[98] Shenoy D, Fu W, Li J, *et al.* Surface functionalization of gold nanoparticles using hetero-bifunctional poly(ethylene glycol) spacer for intracellular tracking and delivery. Int J Nanomedicine 2006; 1(1): 51-7.

[99] Kirchner C, Liedl T, Kudera S, *et al.* Cytotoxicity of colloidal CdSe and CdSe/ZnS nanoparticles. Nano Lett 2005 Feb; 5(2): 331-8.

[100] Chang E, Thekkek N, Yu WW, Colvin VL, Drezek R. Evaluation of quantum dot cytotoxicity based on intracellular uptake. Small 2006 Dec; 2(12): 1412-7.

[101] Hirsch LR, Stafford RJ, Bankson JA, *et al.* Nanoshell-mediated near-infrared thermal therapy of tumors under magnetic resonance guidance. Proc Natl Acad Sci U S A 2003 Nov 11; 100(23): 13549-54.

[102] Loo C, Lin A, Hirsch L, *et al.* Nanoshell-enabled photonics-based imaging and therapy of cancer. Technol Cancer Res Treat 2004 Feb; 3(1): 33-40.

[103] Loo C, Lowery A, Halas N, West J, Drezek R. Immunotargeted nanoshells for integrated cancer imaging and therapy. Nano Lett 2005 Apr; 5(4): 709-11.

[104] Cho SJ, Maysinger D, Jain M, Roder B, Hackbarth S, Winnik FM. Long-term exposure to CdTe quantum dots causes functional impairments in live cells. Langmuir 2007 Feb 13; 23(4): 1974-80.

[105] Lohmann CH, Schwartz Z, Koster G, *et al.* Phagocytosis of wear debris by osteoblasts affects differentiation and local factor production in a manner dependent on particle composition. Biomaterials 2000 Mar; 21(6): 551-61.

[106] Foged C, Brodin B, Frokjaer S, Sundblad A. Particle size and surface charge affect particle uptake by human dendritic cells in an *in vitro* model. Int J Pharm 2005 Jul 25; 298(2): 315-22.

[107] Cherukuri P, Bachilo SM, Litovsky SH, Weisman RB. Near-infrared fluorescence microscopy of single-walled carbon nanotubes in phagocytic cells. J Am Chem Soc 2004 Dec 8; 126(48): 15638-9.

[108] Foucaud L, Wilson MR, Brown DM, Stone V. Measurement of reactive species production by nanoparticles prepared in biologically relevant media. Toxicol Lett 2007 Nov 1; 174(1-3): 1-9.

[109] Gygi SP, Rochon Y, Franza BR, Aebersold R. Correlation between protein and mRNA abundance in yeast. Mol Cell Biol 1999 Mar; 19(3): 1720-30.

[110] Ponti J, Sabbioni E, Munaro B, *et al.* Genotoxicity and morphological transformation induced by cobalt nanoparticles and cobalt chloride: an *in vitro* study in Balb/3T3 mouse fibroblasts. Mutagenesis 2009 Sep; 24(5): 439-45.

[111] L'azou B, Jorly J, On D, *et al. In vitro* effects of nanoparticles on renal cells. Part Fibre Toxicol 2008; 5: 22.

[112] Fletcher JS, Lockyer NP, Vaidyanathan S, Vickerman JC. TOF-SIMS 3D biomolecular imaging of Xenopus laevis oocytes using buckminsterfullerene (C60) primary ions. Anal Chem 2007 Mar 15; 79(6): 2199-206.

[113] Shi H, He X, Yuan Y, Wang K, Liu D. Nanoparticle-Based Biocompatible and Long-Life Marker for Lysosome Labeling and Tracking. Anal Chem 2010 Feb 15.

[114] Choi AO, Brown SE, Szyf M, Maysinger D. Quantum dot-induced epigenetic and genotoxic changes in human breast cancer cells. J Mol Med 2008 Mar; 86(3): 291-302.

[115] Gagne F, Auclair J, Turcotte P, *et al.* Ecotoxicity of CdTe quantum dots to freshwater mussels: impacts on immune system, oxidative stress and genotoxicity. Aquat Toxicol 2008 Feb 18; 86(3): 333-40.

[116] Geys J, Nemmar A, Verbeken E, *et al.* Acute toxicity and prothrombotic effects of quantum dots: impact of surface charge. Environ Health Perspect 2008 Dec; 116(12): 1607-13.

CHAPTER 7

Risks Associated with the Use of Nanomaterials

Sajjad Haider[1*], Nausheen Bukhari[2] and Adnan Haider[3]

[1]*Department of Chemical Engineering, College of Engineering, King Saud University, Riyadh, Saudi Arabia;* [2]*Department of Chemistry, College of Science, King Saud University, Riyadh, Saudi Arabia and* [3]*Department of Chemistry, Kohat University of Science and Technology, Kohat, Pakistan*

> *"It is a mistake for someone to say nanoparticles are safe, and it is a mistake to say nanoparticles are dangerous. They are probably going to be somewhere in the middle. And it will depend very much on the specifics". Quoted by V. Colvin, Director of Center for Biological and Environmental Nanotechnolgy at Rice University, in Technology Review.*

Abstract: Advances in engineering nanostructures with exquisite control of size and shape, their unique properties and broad applications (*e.g.,* probes in ultrasensitive molecular sensing and diagnostic imaging, agents for photodynamic therapy and actuators for drug delivery, triggers for photothermal treatment, and precursors for building solar cells, electronics and light emitting diodes) have made nanotechnology an exciting research area. As the field moves from academic findings to industrial products, concerns have surfaced on the subject of the toxicity of nanostructures. Therefore, it is indispensable to embark on liable development of nanotechnology to develop and use these nanomaterials to meet human and societal needs while making every effort to foresee and alleviate their adverse effects, and unintended consequences. Currently, a complete understanding of the size, shape, composition, and aggregation dependent interactions of nanostructures with biological systems is lacking and thus it is unclear whether the exposure of humans, animals, insects and plants to engineered nanomaterials could produce harmful biological responses. Hence, a new sub-discipline of nanotechnology called nanotoxicology has emerged. There is a keen interest in nanotoxicology research since the processing of nanostructures in biological systems could lead to unpredictable effects. This uncertainty, in combination with the absence of a complete current understanding of the interactions of as-designed nanostructures with biological systems has led to many questions raised by the regulatory agencies and general public to nanotechnology-based products. The focus of this chapter is to update the reader knowledge on the design, synthesis, characterization and in-vitro and in-vivo activities of these nanostructures and define a link between these studies and a better toxicological understanding of nanostructures.

Key Words: Fullerene, Carbon nanotubes, Metals nanoparticles, Metal oxides nanoparticles, Metal alloy nanoparticles, Synthesis, Characterization techniques, Toxicity, *In vivo* toxicity, *In vitro* toxicity.

INTRODUCTION

In the past two decades, the extensive work done on the synthesis and characterization of nanoscale materials were not only given researchers great control over the fabrication of nanomaterials (ranging from 1 to 100 nm), but also unlocked many unique size, shape and composition dependent properties. A variety of engineered nanomaterials, *i.e.,* organic (carbon nanotubes, fullerene derivatives, polymer nanofibers and nano-membranes) and inorganic ((gold (Au), silver (Ag), iron oxide (Fe_3O_4) titanium oxide (TiO_2), silicon oxide (SiO_2) and quantum dots (QDs), *etc.*), and synthetic processes (discharge method, chemical vapour deposition (CVD) and laser ablation (carboneaous nanotubes) electrospinning (polymer nanofibers and nano-membranes), solution precipitation (Fe_3O_4-nanoparticles), microemulsions [1], and polymeric coatings (Au-nanoparticle) [2], chemical reduction/non-aqueous solutions [3-12], template method [13-18], electrochemical and/or ultrasonic-assisted reduction [19-23], photo-induced reduction [24-31], microwave-assisted synthesis [32-37] irradiation reduction [38-41], microemulsion [42-48], biochemical reduction [49-

*****Address correspondence to Sajjad Haider:** Nanofiber Technology Research Lab, Chemical Engineering Department, College of Enginering, King Saud University, P.O. Box 800, Riyadh11421, Saudi Arabia; Tel + 996-1-4675579; E-mail: shaider@ksu.edu.sa

Haseeb Ahmad Khan and Ibrahim Abdulwahid Arif (Eds)

54] (Ag-nanopartiles), hydrolysis and calcinations, reactor flame and furnace synthesis [55] sol-gel method, (TiO2 nanoparticles) [56], hydrolysis and condensation of tetraethylorthosilicate (TEOS) [57] and two-stage hydrolysis in aqueous medium (SiO$_2$-nanoparticles) [58], has emerged, which has opened a new world of creative possibilities. The potential benefits of nanoscale technologies are expected to have substantial impacts on industrial sector *e.g.,* energy (solar cells, fuel cell and energy storage devices), electronics (light emitting diodes, silicon chips), aerospace (development of light weight superior strength materials, radar absorbing coatings, jet and rocket fuel (using aluminium(Al) nano-particles with liquid hydrogen to increase the propulsion energy, *etc.*), medicine (diagnostic imaging, agents for photodynamic therapy (PDT), actuators, drug delivery devices, triggers for photothermal treatment, *etc.*) and social sectors. Serious investment into nanotechnology research started in the period of 1997-2002 (Table **1**), which rose [59, 60] in 2004 to 8.6 billion US$, over 300 nanomaterials summarized in Fig. **1**, accounting for 147 billion US$ were available in market in 2007 [61]. This investment is anticipated to rise to 1 trillion US$ by 2012.

The increase investment in nanoscale technologies is anticipated to have considerable impacts on industrial sectors *e.g.,* energy (solar cells, fuel cell and energy storage devices), electronics (light emitting diodes, silicon chips), aerospace (development of light weight superior strength materials, radar absorbing coatings, jet and rocket fuel (using Al nano-particles with liquid hydrogen to increase the propulsion energy, *etc.*), medicine (diagnostic imaging, agents for photodynamic therapy (PDT), actuators, drug delivery devices, triggers for photothermal treatment, *etc.*) and social sectors. While the speedy transit of novel engineered nanomaterials from scholastic findings to industrial products [65, 66] and their escalated use in the industrial and consumer products is widely made known (Table **2**), potential threats to human health and environment are just beginning to emerge [67].

Table 1: Global R&D expenditure ($M) [62]

Country/ Region	1997	2002
USA	432	604
Western Europe	126	350-400
Japan	120	750
South Korea	0	100 pa (for 10 yrs)
Taiwan	0	70
Australia	0	40
China	0	40
Rest of world	0	270

Table 2: Estamated global production of engineered nanomaterial [70].

Application	Nanomaterial device	Estamated global production (tons/year)		
		2003/04	2010	2020
Structural application	Ceramics, catalysts, film & coating, composites, metal	10	10^3	10^4-10^5
Sinkcare products	Metal oxides(*e.g.,* TiO$_2$and ZnO)	10^3	10^3	10^3
Information and communication technologies	Swnt, nanoelectronics and optoelectronics materials (excludingCMPslurries), organic light emitters, narophosophers	10	10^2	>10^3
Biotechnology	Nanocomposite, encapsulates, target drugdelivery, dagnostic marker, biosensors	<1	1	10
Environmental	Nanofilteration membranes	10	10^2	10^3-10^4

The understanding of the size, shape, composition and aggregation-dependent interactions of the nanostructures with biological systems at present is underexplored, [68] and thus it is hazy whether the exposure of humans, animals, insects and plants to these novel nanostructures could generate harmful biological responses [69]. Therefore, it is indispensable for us to focus on the complete understanding of size, shape, composition and aggregation dependent interactions of the nanostructures with biological system to meet the future human and societal needs [67]. This chapter focuses to update the reader knowledge on the design, synthesis, characterization and in-vitro and in-vivo activities of these nanostructures, and finally to define the link between these studies and a better *in vivo* and *in vitro* toxicological understanding of the nanostructures.

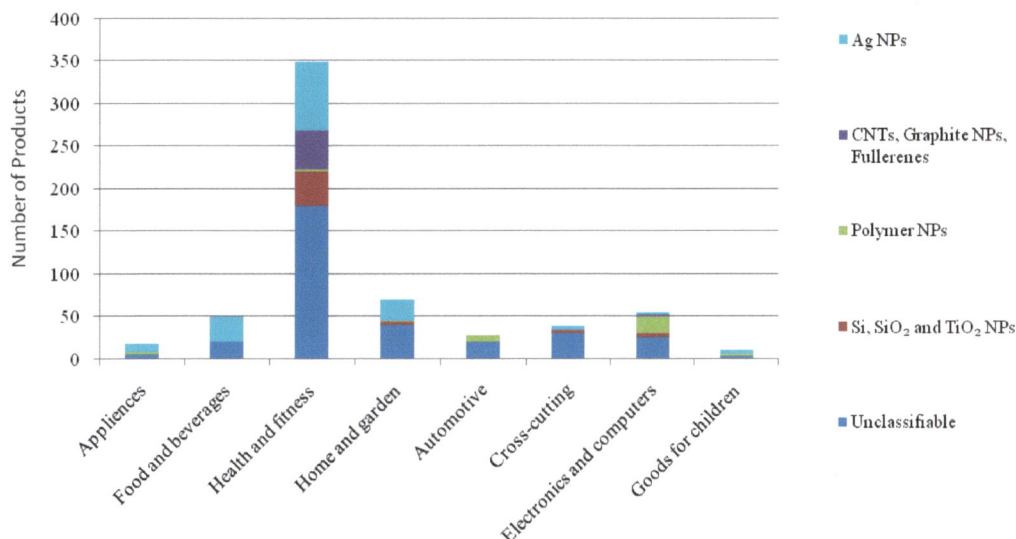

Figure 1: Summary of nanomaterials and their product category [63, 64].

NANOMATERIALS AND THEIR SYNTHESIS

The synthesis and/or fabrication of nanoparticles is as diverse as the materials themselves, fullerenes (allotrope forms of carbon), which exist as hollow spheres(bucky balls), ellipsoids, and nanotubes (Multi-walled carbon nanotubes (MWCNTs) and single-walled carbon nanotubes (SWCNTs)) occurs naturally as combustion products and fabricated by vaporization of graphite by resistive heating [71], combustion of simple hydrocarbons in fuel-rich flames [72] and UV laser irradiation of geodesic polyarenes [73, 74]. Four advances in the UV laser irradiation of geodesic polyarenes made this synthesis feasible; curvature was provisionally induced in polyarenes *via* flash-vacuum pyrolysis, radical-initiated C(aryl)-C(aryl) coupling reactions were designed to interdict the distorted conformations, facile 1, 2-hydrogen shifts were exploited to limit challenging synthetic transformations, and cyclo-dehydrogenation cascades stitched the developing p-system together once curvature was induced [75]. Significant synthetic challenges still to be over come to prepare higher order fullerenes, ^{13}C-labeled fullerenes, heterofullerenes, and azafullerenes, *etc.* Carbon nanotubes (CNTs) were discovered in 1991 in cathode deposits *via* arc evaporation of graphite [76]. Shortly after their discovery, CNTs were also isolated as an end product of hydrocarbons (ethylene or acetylene) pyrolysis over iron, cobalt, and other dispersed metals [75-79]. The presence of metals greatly influences the size profile of CNTs [80]. MWCNTs were prepared by pyrolysis of metallocenes (ferrocene, cobaltocene, and nickelocene) under reducing conditions; metallocene acts both as a source carbon and metal [81]. Pyrolysis of nickelocene in the presence of benzene at $1100^{\circ}C$ yields primarily MWCNTs. In contrast, the use of acetylene in nickelocene pyrolysis primarily yielded SWCNTs presumably due to the smaller number of carbon atoms per molecule [76]. SWNTs were also prepared in a related approach using dilute hydrocarbon-organometallic mixtures [82, 83]. Both CNTs and bucky balls due to their high aspect ratio, high strength, electrical conductivity, electron affinity, structure, and versatility have created potential academic and commercial interest [84]. Metals and metal oxides and metal composites (alloy) *i.e.,* Au, Ag, TiO_2, SiO_2, Fe_3O_4 and QDs, respectively, are functional materials having unique optical, electrical and magnetic properties properties [85, 86]. Au nanoparticles are commonly prepared *via* chemical reduction of

Au salts in aqueous, organic, or mixed solvent systems in the presence of a citrate [83] and thiol-containing organic groups [1] stabilizers (the stabilizers bind to the surface and prevent aggregation *via* favorable cross-linking and charge properties), microemulsions [87], and polymeric coatings [88], *etc.* Ag came into use even before Neolithic revolution; Greeks used Ag for cooking and keeping the drinking water safe [89]. Many diverse approaches have been use to synthesis Ag nanoparticles, these include chemical reduction of Ag ions in aqueous solutions [3-10] or non-aqueous solutions [11-13], template method [14-19], electrochemical or ultrasonic-assisted reduction [19-23], photo-induced reduction [24-31], microwave-assisted synthesis [32-37], irradiation reduction [38-41] microemulsion [42-48] and biochemical reduction [49-54], *etc.* TiO_2 nanoparticles were synthesized by a variety of approaches, some of the recently reported approaches are; low-pressure spray pyrolysis (LPSP) organic precursors [90], hydrolysis and condensation of titanium tetra ethoxide (TEOT) and vaporized water using a continuous aging tube reactor [91], hydrolysis of titanium isopropoxide [92], reactive direct-current magnetron sputtering [93], laser pyrolysis of titanium tetrachloride-based gas-phase mixtures [94], and sol-gel technique modified by incorporation of ultrasound as a reaction aid, *etc.* [95]. SiO_2 were synthesized by oxidation of tetraethyl-orthosilicate (TEOS) in the bench-scale diffusion flame rector [96], hydrolysis and condensation of tetraethylorthosilicate (TEOS) using continuous microwave process [97], two-stage hydrolysis of silicon powder in aqueous medium [98], ultrasonication by sol-gel process [99] and flame pyrolysis. [100]. Fe_3O_4 is investigated the most in biomedical techniques; due to its superior biocompatibility compared other magnetic materials based on oxides or on pure metals [101].

Numerous types of iron oxides exist in nature and can be prepared in the laboratory, but maghemite (γ-Fe_2O_3) and magnetite ((Fe_3O_4) Fig. **2**) with sufficiently high magnetic moments, chemical stability in physiological conditions and low toxicity, simple and economical synthetic procedures fulfill the necessary requirements for biomedical applications. A number of synthetic procedures are in practice to synthesize Fe_3O_4 nanoparticles. The simplest, most economical and environmentally-friendly procedure is based on the co-precipitation method, which involves the precipitation of Fe^{2+} and Fe^{3+} ions in basic aqueous media. Apart from co-precipitation method, these are also prepared *via in-situ* synthesis at room temperature in the presence of modifier. Highly dispersed α-Fe_2O_3 nanoparticles were prepared at atmospheric pressure, low temperature, and at an ultra-dense reagent concentration by titrating an aqueous ammonia solution into a dense iron oleate/toluene mixture [101].

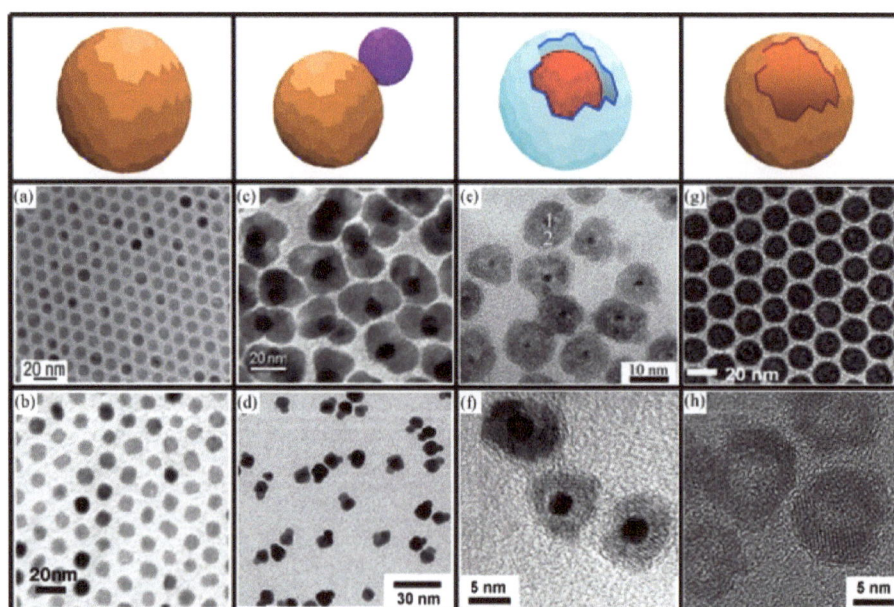

Figure 2: Various iron-based inorganic nanostructures are now available, *e.g.,* (a) Fe_3O_4 nanoparticles, (b) FePt alloy, (c) heterodimers based on Fe_3O_4-Au (c) or (d) FePt-Au, (e) yolk–shell nanostructures composed by FePt core and CoS_2 shell or (f) Pt–Fe_2O_3, hollow nanocrystals based on (g) Fe_3O_4 or (h) γ-Fe_2O_3 [102].

QDs have a metalloid crystalline core (*e.g.,* CdSe) and a shell (*e.g.,* ZnS) that shields the core. The core consists of a variety of metal complexes such as semiconductors, noble metals, and magnetic transition metal and the shell is formed on the metalloid core during synthesis in organic solvents, which makes QDs hydrophobic and therefore limits their use in biological applications. To make QDs biocompatible, secondary coatings are added onto them to improve their water solubility. QDs display narrow fluorescence band due the quantum constraints imposed on electrons by the finite their size [103], which makes QDs advantageous as optimal fluorophore for in-vivo biomedical imaging and for targeting specific cells after conjugation with specific bioactive moieties [104, 105].

Fluorescent QDs conjugated with bioactive moieties (*e.g.,* antibodies, receptor ligands) were used to label neoplastic cells, peroxisomes, DNA, and cell membrane receptors. Bio-conjugated QDs are also being explored as tools for site-specific gene and drug delivery and are among the most promising candidates for a variety of information and visual technologies. They are currently used for the creation of advance flat panel LED (light-emitting diode) displays and may be employed for ultrahigh-density data storage and quantum information processing [106].

NANOMATERIALS DESIGN AND CHARACTERIZATION TECHNIQUES

Nanomaterials have unique properties (*e.g.,* high aspect ratio, high strength and high surface area to volume ratio) relative to bulk materials, which may bestow on them unique mechanisms of toxicity. Toxicity in particular has been considered to originate from nanomaterials size, aggregation, surface area, composition and shape [107]. Size plays a vital role in modeling the behavior of the body to respond to, distribute, and eliminate materials [108]. Particle size affects the mode of endocytosis, cellular uptake, and the efficiency of particle processing in the endocytic pathway [107]. Decreasing the size leads to an exponenional increase in surface area relative to volume, thus making the nanomaterials surface more reactive on itself (aggregation) and to its surrounding environment (biological components).

Chemical composition at the surface of nanomaterials will define their chemical interactions because the surface is in direct contact with the body whereas the limited bulk volume is hidden [108]. Many nanomaterials are functionalized on the surface to make them more biocompatible and potentially interactive with biological components, alter biological function, and penetrate into certain cells. Degradability of the material is another significant component of nanomaterials toxicity causing acute and long-term toxicity. Non-degradable nanomaterials can accumulate in organs and also intra-cellularly, where they can cause detrimental effects to the cell, similar to that of lysosomal storage diseases [109]. In contrast, biodegradable nanomaterials can lead to unpredicted toxicity due to unexpected toxic degradation products [110]. Nanomaterials may also contain transition metals (*e.g.,* quantum dots) or other compounds with known toxicity that are "masked" for instance by functionalization. Degradation may release toxins to the biological environment, leading to free radical formation and resulting in cellular damage [107, 110].

Therefore prior to evaluate the biological responses of nanomaterials to real scientific validity of any degree in-vitro or in-vivo testing, it is imperative to exercise quality control by thoroughly screening their inherent aqueous stability, aggregation, flocculation, procedures for the removal of unwanted material and batch variability. Most of the exceptional properties of nanocsale material originate from their dimension (nanomaterials have one facet less than 100 nm). Due to high dispersion and elevated surface energy, nanoparticle aggregation is common in biological media and at nanoscale it is exceptionally difficult to differentiate aggregation, predominantly in biological environment, which could put forth prominent effect on nano-specific interaction with the cells as well as tissues. Aggregation pre-dominantly affects homogeneity, colloidal stability, optical and electronic, cell and/or bacterial uptake properties in biological solution. Therefore, it is imperative to carry out first sizing of particle and aggregation stability assays in biological solution before studying their homogeneity and colloidal stability, optical and electronic properties, and cell and/or bacterial uptake. The methods currently used for particle sizing and determination of nanomaterials aggregation are; transmission electron microscopy (TEM), scanning electron microscopy (SEM), optical spectroscopy (UV-vis), dynamic light scattering (DLS), and fluorescence polarization. The above mentioned methods do not prove crucial in complex nanomaterials systems; therefore, these are best suited for pre-experimental characterization of nanomaterials. TEM is a entrenched direct electron imaging technique under ultrahigh vacuum (UHV) conditions for studying shape, morphology, particle size distributions and aggregation of nonmetallic and in particular of metallic nanomaterials [111-114], though in the former case emission field transmission electron microscope (EFTEM) might be used due to the energy filtering discrepancy [115, 116] (Fig. **3**). However having said this, the information provided by TEM in *ex-situ* particle aggregation doesn't not necessarily represent the *in-situ* aggregation states due to the artifacts (*e.g.,* aggregation as a result of the ionic potency and surface tension), which are produced in the sample during its preparation. Flash freezing and desiccation techniques could be used to eliminate some of these artifacts; however it is pretty tiresome and needs a lot of practice. For this very reason TEM solution aggregation information should be corroborated by other methods (*i.e.,* particle sizing and zeta-potential and gel electrophoresis) [113]. Sometimes, TEM are also equipped with energy dispersive X-ray spectroscopy devices (EDS) (for determining elemental content), X-ray absorption spectroscopy (for determining three dimensional structures) and atomic force microscopy (AFM) (for measuring three dimensions morphology of the surface). Despite its limitation TEM is considered a powerful tool to differentiate crystalline particulate chemistry and to examine material crystal habit/aggregation (*e.g.,* nano-fullerene aggregate samples were distinguished from nano-fullerene crystalline samples in resin-fixed and freeze-dried cells examined by TEM micro-diffraction) [113-117]. SEM is another technique, which is used pre-dominantly for imaging the surface texture of materials, however; it can also be used for particle sizing and aggregation. Conventional SEM requires UHV environment and dehydrated samples (dry samples), which lead to uncertainty in getting accurate data in case of wet samples [113]. The problem has been overcome and nowadays the liquid (aqueous) surface and substances close to the surface can be imaged in *in-situ* material characterization while maintaining UHV conditions around the electron gun with commercially available emission scanning electron microscopy (ESEM) (Fig. **3**). Furthermore, an alteration in the standard protocols (by utilizing Peltier element to control evaporation) and transmission mode of ESEM extended its abilities to image even emulsions and particle suspensions [118].

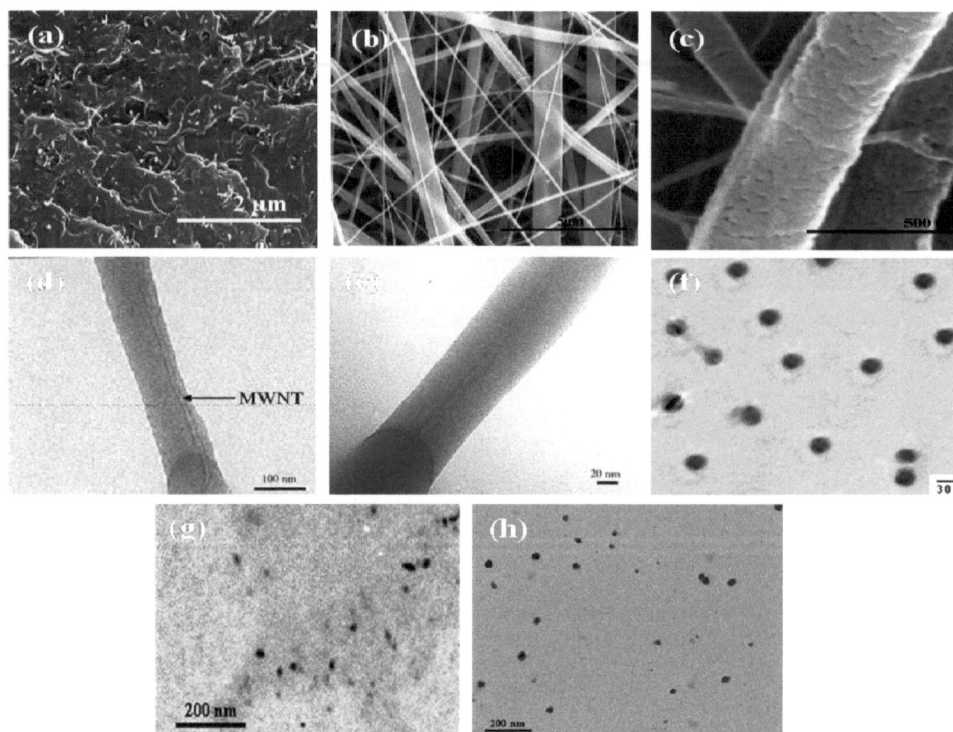

Figure 3: SEM micrographs; (a) MWCNTs, (b) chitosan nanofibers, (c) Freez-dried chitosan nanofibers (d) TEM micrographs of functionalized MWCNTs/caprolactone nanocomposite nanofiber, (e) functionalized MWCNTs/caprolactone nanocomposite nanofiber at magnification, (f) Ag nanoparticles, (g) Oleic acid coated magnetic nanoparticles and (h) maleic anhydride-co-b-cyclodextrine coated magnetic nanoparticles [119-123].

Surface plasmon resonance (SPR) is a technique, which is used for sizing and studying of nanoparticles colloidal aggregation. This technique mainly depends on shape, diameter, adsorbed adsorbates and the distance between plasmonic particles. According to Mie theory, certain metal nanoparticles and nanorods (pure Pb, In, Hg, Sn, Cd Ag, and Au) displayed absorption and scattering of light *via* excitation of the metal's plasmon (communal oscillations of the free electron density) band by incident or scattering of photons of the proper wavelengths. This absorption and scattering of light is dependent on the size (of the nanoparticle), metallic electrons, lattice structures, and interband electron physics and excitation energies. While measuring the size of the metal nanpaprticles, plasmon peak absorbance give red shift (moves to high wavelengths) when the average diameter of the particle increases. The absorbance shift (red) differentiates the adsorption of contaminants and stabilizing layers onto the nanoparticle surface from that of the clean material only by nm [124]. At present Ag/Au nanoparticles are commonly sized by measuring the extinction wavelength(s) of incident light. Pt, Pd and Cu also have plasmonic optical properties when protected from significant oxides contaminants. Furthermore, the inter-particle distances when smaller than average particle diameter will result in a shift in sample absorbance, which could be used as a particle aggregation indicator (*e.g.,* for aggregated Ag colloidal solution, the color changes from red to blue). However, in biological environment the effect of adsorption onto particle surface and the resulting aggregation (due to the non-specific adsorption) are hard to differentiate [113]. Dynamic light scattering (DLS) has been used extensively in nanotechnology for quick and precise measurement of particle (Fig. **4a, b**) hydrodynamic size in solution, polydispersities and aggregation effects of samples [123, 125]. However, nano-sizing by DLS not only need good level of expertise to calibrate and control the instrument, but also sufficient knowledge and information of reagent purities and optical data modeling algorithms to forecast sizing consistent with many assumptions of optical scattering in solution. Apart from this, DLS data is also very sensitivity to traces of contaminants *e.g.,* increasing salt decreases colloid stability, whereas non-polar adsorbate increases colloid stability through steric stabilization. These unwanted adsorbates, solutes and aggregates should be broken down in solution through bath sonication followed by dialysis or centrifugation before analysis [111]. Care must be taken as conventional sonication treatments (bath and probe) are difficult to regulate (density, dose, power, and local heating) and sometimes have unnecessary side effects (such as the addition of metal particles typically shed from the probe tip during sonication treatment and oxidation of active surface) [126]. Hence many particle sizing such as spanning metals, metal oxides and polymers use this method blindly for various sample type, with default scattering models and curve fitting assumptions, which are neither justified and nor validated [113].

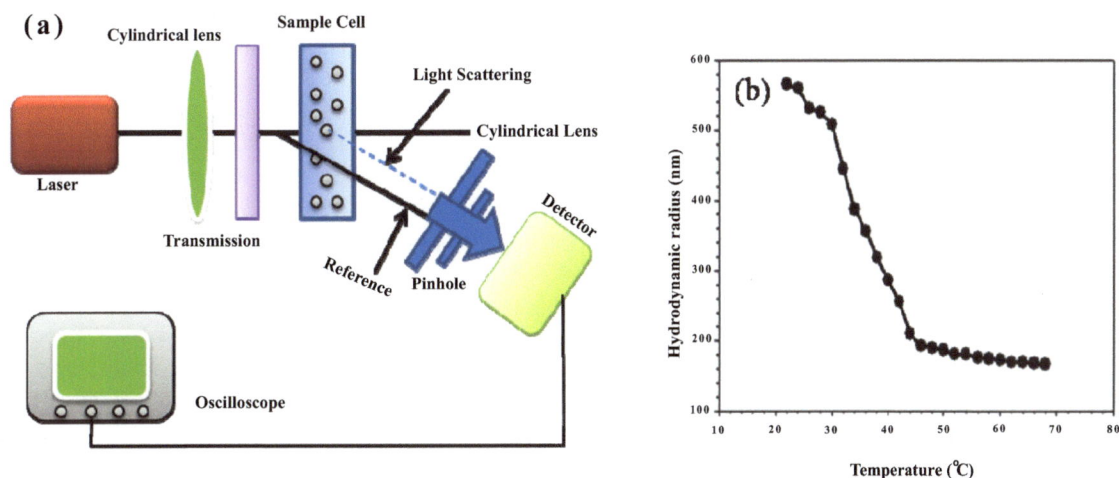

Figure 4: (a) Schematic of the optical setup of dynamic light scattering (DLS) measurement sing a transmission grating and (b) Measurement of hydrodynamic radii of the poly (NIPAAm)-MNPs dispersed in distilled Water *via* DLS. [123].

To carry out accurate particle sizing, DLS studies should therefore be designed carefully by doing proper sample calibration with sizing standards under conditions pertinent to the experiment and thorough descriptions of the data analyses employed [113]. Various standards are available now to aid investigators in the sizing of the nanoparticles, these include the American Society for Testing and Materials (ASTM) standard for the measurement of nanomaterials particle size distribution in suspension by photon correlation spectroscopy (ASTM E2490-08) and national institute of standards and Technology (NIST) "Au standard" (NIST RM8011, NIST RM8012 and NIST RM8013).

However, in case of the biological environment, which is sensitive to ionic strength, polymer, surfactant, peptide, and proteins investigators must look for more accurate methods for obtaining reproducible data. Time-resolved fluorescence polarization anisotropy (TRFPA) is another technique, which has been used for particle sizing. According to Stokes-Einstein-Debye rotational equation of particle motion, TRFPA correlates fluorescence polarization decay time to fluor or particle hydrodynamic radius. This is highly sensitive method and could measure particles from 1 to 10 nm with 0.1 nm resolution. In assay environment, TRFPA utilizes sub-nanosecond-resolution laser pulses and detectors to excite fluors. The fluorescence induced by optical excitation can be illustrious in free-floating versus receptor or surface-bound states by emission anisotropy. The decay of fluorescence polarization distinguishes particle or fluor binding to receptors from assay components. Having said this TRFPA has not up till now established its wide application in nanomaterial characterization [113]. Other techniques, which have been used for routine sizing of nanomaterials are Large angle X-ray diffraction (XRD) [123, 126], Multi-angle laser light scattering (MALLS) in combination with UV-vis spectroscopy, small angle X-ray scattering (SAXS) and small angle neutron scattering (SANS). Addionally supportive techniques such as inductively coupled plasma-mass spectroscopy (ICP-MS), inductively coupled plasma-atomic emission spectroscopy (ICPAES) and thermogravimetric analysis (TGA) are also in practice to measure traces of metals, Au (atoms) concentration in solution, and weight loss study, respectively.

NANOMATERIALS: THE POTENTIAL CAUSE OF TOXICITY

Nanotechnology is at the convergence of engineering, chemistry, physics and molecular biology, leading to the development of structures and devices with novel functional properties, which can be used in a various sectors of human life as discussed previously, but in particular will have potential benefits energy (solar cells, fuel cell and energy storage devices), electronics (light emitting diodes, silicon chips), aerospace (development of light weight superior strength materials, radar absorbing coatings, jet and rocket fuel (using Al nano-particles with liquid hydrogen to increase the propulsion energy, *etc.*), medicine (diagnostic imaging, agents for photodynamic therapy (PDT), actuators, drug delivery devices, triggers for photothermal treatment, *etc.*) and social sectors. While the speedy transit of novel engineered nanomaterial from academic findings to industrial products [68, 69] and their escalated use in industrial and consumer products are widely made known (Table **2**), potential threats to human health and environment are just beginning to emerge [67]. Several aspects of societal implications of nanoscience and nanotechnology in developing countries have been recently discussed. Timely assessment of nanomaterial toxicity will provide this critical data, improve public trust of the nanotechnology industry, assist regulators in determining environmental and health risks of commercial nanomaterials, and provide industry with information to direct the development of safer nanomaterials and products [67]. The use of nanomaterials in biological system is very complex due to their unexplored risks to human health, thus while conducting quantitative studies on the uptake of nanomaterials into biological systems, the simultaneous agglomeration, sedimentation and diffusion at physiologically relevant concentrations should be taken into account. Furthermore, it is also imperative to assess the ecological risk, to understand environmental implications of nanomaterials. The fate of nanomaterials in ecosystem (aqueous environment) is controlled by many biotic/abiotic processes such as solubility, dispersability, interactions between the nanomaterials and natural/anthropogenic chemicals. Outburst of nanomaterial research will certainly pump a lot of to the environment, which will ultimately lead to nanomaterials pollution. To date, there are no detailed studies on the transport mechanism, biodegradation or association of nanomaterials with biological materials that may eliminate nanomaterials. Processes that control transport and removal of nanomaterials in water and wastewater are yet to be investigated to understand the fate of nanomaterials. Studies on the effect of nanomaterials on plants and microbes are also rare. Alumina nanoparticles have been frequently employed in UV protections products (scratch-resistant transparent coatings and sunscreen lotions). However, a recent study reported that alumina nanoparticles led to phytotoxicity *i.e.,* retarding the root growth of the corn, cucumber, soybean, cabbage and carrot. In another study, toxicity of fullerene-C_{60} in Daphnia and

Pimephales elevated lipid peroxidation (LPO) in brain, significantly increased LPO in gill, and resulted in significant increase in expression of genes related to the inflammatory response and metabolism (CYP2 family). Studies on the effect of nanomaterials on human health have lately gained momentum; about 1.5-2 million deaths per year worldwide could be ascribed to indoor air pollution. Nanomaterials component of particulate air is highly potent and likely to be accountable for adverse health effects *e.g.*, carbon nanomaterials when inhaled are capable of rapid translocation into the circulation and could also remain within the lung upto 6h after inhalation, without passing to systemic circulation. Cytotoxicity associated with nanoparticles exposure is somewhat particle specific (Table **3**).

Table 3: Toxicity of selected nanomaterials [127].

Fullerene C_{60} water suspension	Antibacterial; cytotoxic to human cell lines; taken Up by human keratinocytes; stabilizes proteins
C_{60} encapsulation in poly(vinyl-pyrrolidone), cyclodextrine, or poly(ethylene glycol)	Demages eukaryotic cell lines; antibacterial
Hydroxylated fullerene	Oxidative eukaryotic cell damage
Carboxyfullerene (malonic acid derivatives)	Bactericidal for gram-positive bacteria; cytotoxic to human cell lines
Fullerene derivatives with pyrrolidine groups	Antibacterial; inhibits cancer cell proliferation; cleave plasmid DNA
Other alkane derivatives of C_{60}	Anti mutagenic; cytotoxic; induces DNA damage in plasmids; inhibits protein folding; antibacterial; accumulates in rats livers
Metallofullerene	Accumulates in rats livers
Silicone dioxide (SiO_2)	Pulmonary inflammation in rats
Titanium dioxide (TiO_2) (Anatase)	Antibacterial; pulmonary inflammation in rodents
Zinic oxide (ZnO)	Antibacterial (micrometer scale); pulmonary effects in animals and humans

The *in-vitro* assay conducted on the spermatogonial stem cell line in the male germline Ag nanoparticles were found the most toxic while molybdenum trioxide (MoO_3) were least toxic.The toxicity of Cu nanoparticles was assessed *in vivo* based on LD_{50}, morphological changes, pathological examinations and blood biochemical indices of experimental mice. Serious nanotoxicity was accredited to high specific surface area, ultrahigh reactivity and high consumption of H^+, having said this, not all nanoparticles are dangerous. The toxicology and biodynamics of SiO_2 nanoparticles investigated in a mice model revealed that nanoparticles nanoparticles were not toxic and can be used *in-vivo*.

Results generated from the procedure developed by an insurance company for the purpose of calculating insurance premium for chemical manufacturers revealed that the relative environmental risk from manufacturing SWCNTs, C_{60}, quantum dots, alumoxane nanoparticles and TiO_2 was comparatively lower than common industrial manufacturing processes. This study should not be misunderstood to endorse the manufacture of these nanomaterials without their detailed evaluation of environmental and human risks. Besides workers others then manufacturing wing (*e.g.,* occupational health nurses), who get exposed to nanoparticles should be aware of the potential risks and possible means to avoid health risks. There is a need to classify exact regulatory regimes to protect personnel involved in the production and use of nanoparticles for cosmetic, medical and agricultural purposes. At the current speed of research in the nanotechnology and the risk associated with it, we may need several years or decades to clearly establish the health and environmental risks from engineered nano-scale particles [67].

ASSESSMENT OF NANOMATERIALS TOXICITY (*IN-VIVO*)

Nanomaterials have shown dose, size and functional group dependent *in-vivo* toxicity *e.g.,* SWCNTs demonstrated dose-dependent effects of interstitial inflammation and lesions in mice and rats (at a concentration of 0-0.5 $mg \cdot kg^{-1}$ and exposure time of 7 to 90 days), pulmonary grandulomas in rats after exposure to SWCNT soot (at concentrations of 1 and 5 $mg \cdot kg^{-1}$ and exposure time of 24 h to 3 months), however; lateral effects were not dosage dependent. Ecotoxicity tests of SWCNTs (dispersed in sodium dodecyl sulphate (SDS) *via* sonication) carried out on juvenile rainbow trout (at concentrations 0.1, 0.25 and 0.5 $mg \cdot L^{-1}$ and exposure time of 24 h to 10 days) showed not only a dose-dependent rise in ventilation rate, gill pathologies (oedema, altered mucocytes, hyperplasia), and mucus secretion with SWCNTs precipitation on the gill mucus, but also noteworthy dose-dependent reduction in thiobarbituric acid reactive substances (TBARS), especially in the gill, brain and liver, which is an indication of oxidative stress. On other hand, MWCNTs exhibited morphology based acute toxicity in rats with LD_{90} of 5 mg·kg^{-1} *e.g.,* long MWCNTs caused noteworthy inflammation and tissue damage in mice, while shorter MWCNTs caused less inflammation. Furthermore, water-soluble components of MWCNTs do not produce strong inflammatory effects in mice. Most studies on the toxicological effects of C_{60} suggest that these materials tend to induce oxidative stress in living organisms. Functionalized C_{60} (f- C_{60}) *e.g.,* hydroxylated C_{60} (C_{60} (OH) caused acute oxidative stress in living organisms. *e.g.,* in male mongrel dog significant increase in lipid peroxidation (LP) products was observed after intravenous administration of C_{60} (OH) at concentration of 1 mg. kg^{-1}. In the same way, elevated LP was also observed in the brain and gills of daphnia magna after exposure to hydroxylated (C_{60} (OH)) and tetrahydrofuran (THF)-dissolved C_{60} (THF was shown to have no contribution to the effect mentioned) and LD_{50} of 600 mg·kg^{-1} of polyalkylsulfonated-C_{60} in female rats after intraperitoneal administration (at a concentration of 0-2500 mg·kg^{-1} and exposure time of up to two weeks). Metal nanoparticles were found to induce more severe lung toxicity in mice than same bulk materials. The effects of Zn on humans were studied and after 2 h of exposure to 5 mg·m^{-3} concentration of Zn nanoparticles, the exposed individuals started feeling, sore throat, chest tightness, headache, fever and chills whereas adverse effects were found when the same test was repeated in three trials, 2 h each, however no effect was observed at lower concentration (*i.e.,* 500 μg·m^{-3}). Environmental exposure to Zn nanoparticles could cause pulmonary (lung) inflammatory response in mice. Zn nanoparticles were also found to cause severe symptoms of lethargy, anorexia, vomiting diarrhea, loss of body weight and even death in mice when administered gastro intestinally whereas limited effect were observed for micro-scale Zn at equal concentrations. These studies suggest that Zn nanoparticles toxicity is concentration-dependent and the most probable uptake path is through the respiratory system. Al nanoparticles administration at concentration of 2 mg·mL^{-1} for 24 h to Zea mays (corn), Glycine max (soybean), Brassica oleracea (cabbage), and Daucus carota (carrot), inhibited their growth. Metal oxide nanoparticles such as TiO_2 and SiO_2 also showed size dependent toxicity *e.g.,* smaller TiO_2 and SiO_2 naoparticles caused severe pulmonary damage in mice and stronger lung inflammation in rats, respectively as compared to larger their ones. Administration of single-dose intravenous bolus of FeO naoparticles (at concentration of 20 and 200 mg.kg^{-1}) caused hypoactivity, ataxia, emesis, exophthalmos, salivation, lacrimation, discolored and mucoid feces, injected sclera, and yellow eyes in dogs and a significant increase in fetal skeletal malformations in rats and rabbits [64].

ASSESSMENT OF NANOMATERIALS TOXICITY (*IN-VITRO*)

A number of cytotoxicity studies with SWCNTs and MWNCTs are reported in the literature. Unrefined SWCNTs ((iron (Fe) containing) at concentration of 0.6 to 0.24 g·mL^{-1} and exposure time of 2 to 18 h) exerted oxidative stress and cellular toxicity in human epidermal keratinocytes, refined SWCNTs (at concentration of 0.8 and 200 μg·mL^{-1}) inhibited cell proliferation and decreased cell adhesive-ability in human embryo kidney cells. Surface functionalization of SWCNTs plays an important role in their cytotoxicity towards human dermal fibroblasts. MWCNTs were found more cytotoxic towards Jurkat T leukemia cells after oxidation, Pristine MWCNTs (in concentration of 0.1, 0.2, and 0.4 mg·mL^{-1} and exposure time of 1 to 48 h) decreased the viability of human osteoblastic lines and human epidermal keratinocytes. The cytotoxicity of commercially obtained MWCNTs in bacterial systems before and after physicochemical modification were compared, the highest toxicity was observed when the nanotubes were short, uncapped, and well dispersed in solution. From the above discussion, it could be concluded that there is need for careful documentation of the physical and chemical characteristics of CNTs, when reporting *in-vitro* toxicity.

Contrasting reports about the toxicity caused by fullerenes (C_{60}) are available in the literature *e.g.,* in one report reduced viability of bovine and human alveolar macrophages and increased levels of cytokine mediators of inflammation (*i.e.,* IL-6, IL-8 and TNF) was observed on exposure to sonicated C_{60}, while in another, no toxicity of C_{60} and raw soot were found. The reason behind the discrepancy between the results of two can be attributed to the fact that the later study used TEM to image the distributions of the C_{60} within the macrophages, while former used a viability assay, based on metabolic activity as primary parameter. Studies on the effects of nanoparticles on alveolar macrophages are very important because the alveolar macrophages are the first line of cellular defense against respiratory pathogens. The dose-dependent cytotoxicity of C_{60} (OH) (1-100 $\mu g \cdot mL^{-1}$ for 24 h) resulted a decrease in the cell density and lactate dehydrogenase (LDH) release in human umbilical vein endothelial cells cavity (a sign of increase in non-viable cell numbers). Furthermore dose-dependent causes decrease in the viability of human epidermeal keratinocytes after exposure to C_{60}-phenylalanine (no contribution to the effect was attributed to the phenylalanine groups). The toxicity of QDs was found to be influenced by composition, size, surface charge and coating. Cadmium selenide (CdSe)/zinc sulfide (ZnS) QDs (*i.e.,* CdSe QDs in a ZnS matrix), coated with dihydrolipoic acid (DHLA) had no effect on mammalian cells, while CdSe/ZnS QDs coated with albumin showed undesirable effects on mouse lymphocytes. QDs also showed size, light and temperature dependent cytoxicity *e.g.,* under the same conditions smaller and positively (2.2 ± 0.1 nm) charged QDs exhibit stronger cytotoxicity than larger (5.2 ± 0.1 nm). Cytotoxicity of QDs is also subjective to light and temperature exposure. About 56% damaged deoxy ribonucleic acid (DNA) was observed when exposure to CdSe/ZnS together with ultraviolet (UV) light as compared to only 29% when exposed to CdSe/Zn in the absence of UV light. Similarly CdSe/(cadmium sulfide) CdS (*i.e.,* CdSe QDs in a CdS matrix) were found toxic to cancer cells at 37 °C, but showed no toxicity at 4°C at all. Nano-particulate Ag is an effective bactericide against S. epidermidis, it effectively kills E. coli bacteria too. Increase in the production of LDH levels (an indicator of inflammation) in immortalized rat lung epithelial cells was observed, when these are exposed to 520 $\mu g \cdot cm^{-2}$ Zn nanoparticls for 1 h. Metal oxide nanoparticles such Anatase TiO_2, killed human dermal fibroblast (HDF) cells at LC_{50} of 3.6 $\mu g/mL$ and decreased the viability of human lymphoblastoid cells at concentration of 0-130 $\mu g \cdot mL^{-1}$ when exposed to 6-48 h, SiO_2 nanoparticles considerably inhibit replication and transcription in human epithelial HEp-2 cells (at a concentration of 25 $\mu g \cdot mL^{-1}$ and exposure time of 24 h) and Fe_3O_4 nanoparticles coated with dextran decreases the viability of human monocyte macrophages. [64].

REFERENCES

[1] Pileni MP, Petit C, Lixon P. *In situ* syntheses of silver nanocluster in AOT reverse micelles. J Phys Chem 1993; 97 (49): 12974-83.

[2] Suslick KS, Fang M, Hyeon TJ. Sonochemical synthesis of iron colloids. J Am Chem Soc 1996; 118 (47): 11960-61.

[3] Leopold N, Lendl B. New method for fast preparation of highly surface-enhanced raman scattering (sers) active silver colloids at room temperature by reduction of silver nitrate with hydroxylamine hydrochloride. J Phys Chem B 2003; 107 (24): 5723-27.

[4] Caswell KK, Bender CM, Murphy CJ. Seedless, Surfactant less wet chemical synthesis of silver nanowires. Nano Lett 2003; 3 (5): 667-9.

[5] Pillai ZS. Kamat PV. What factors control the size and shape of silver nanoparticles in the citrate ion reduction method. J Phys Chem B 2004; 108 (3): 945-51.

[6] Yin Y, Li ZY, Zhong Z, Gates B, Xia Y, Venkateswaran S. Synthesis and characterization of stable aqueous dispersions of silver nanoparticles through the tollens process. J Mater Chem 2002; 12 (3): 522-7.

[7] Zhu YC, Ji M, Zheng HG, Li Y, Yang ZP, Qian YT. Seed-mediated synthesis of silver with skeleton structures. Mater Lett 2004; 58 (6):1121-26.

[8] Chaki N.K, Sharma J, Mandle AB, Mulla IS, Pasricha R, Vijayamohanan K. Size dependent redox behavior of monolayer protected silver nanoparticles (2-7 nm) in aqueous medium. Phys Chem Chem Phys 2004; 6 (6):1304-09.

[9] Sun YG, Mayers B, Herricks T, Xia Y. Polyol synthesis of uniform silver nanowires: a plausible growth mechanism and the supporting evidence. Nano Lett 2003; 3 (7): 955-60.

[10] Sun Y, Xia Y. Shape-controlled synthesis of gold and silver nanoparticles. Science 2002; 13 (5601): 2176-79.

[11] Chen DH, Huang YW. Spontaneous Formation of Ag nanoparticles in dimethylacetamide solution of poly (ethylene glycol). J Colloid Interface Sci 2002; 255(2): 299-302.

[12] Wang XQ, Itoh H, Naka K, Chujo Y. Tetrathiafulvalene-assisted formation of silver dendritic nanostructures in acetonitrile. Langmuir 2003; 19(15): 6242-46.

[13] Faure C, Derre A, Neri W. Spontaneous formation of silver nanoparticles in multilamellar vesicles. J Phys Chem B 2003; 107(20): 4738-46.

[14] Chen S, Carroll DL. Synthesis and characterization of truncated triangular silver nanoplates. Nano Lett 2002; 2 (9): 1003-07.

[15] Mandal S, Rautaray D, Sastry M. Ag^+–Keggin ion colloidal particles as novel templates for the growth of silver nanoparticle assemblies. J Mater Chem 2003; 13 (12); 3002-05.

[16] Malandrino G, Finocchiaro ST, Fragala IL. Silver nanowires by a sonoself- reduction template process. J Mater Chem 2004; 14 (18): 2726-28.

[17] Behrens S, Wu J, Habicht W, Unger E. Silver nanoparticle and nanowire formation by microtule templates. Chem Mater 2004; 16 (16): 3085-90.

[18] Morley KS, Marr PC, Webb PB, *et al.* Clean preparation of nanoparticulate metals in porous supports: a supercritical route. J Mater Chem 2002; 12(6): 1898-1905.

[19] Johans C, Clohessy J, Fantini S, Kontturi K, Cunnane VJ, Electrosynthesis of polyphenylpyrrole coated silver particles at a liquid–liquid interface. Electrochem Commun 2002; 4 (3): 227-30.

[20] Zhang YH, Chen F, Zhuang JH, *et al.* Synthesis of silver nanoparticles *via* electrochemical reduction on compact zeolite film modified electrodes. Chem Commun 2002; (23): 2814-15.

[21] Ma HY, Yin BS, Wang SY, *et al.* Synthesis of silver and gold nanoparticles by a novel electrochemical method. Chem Phys Chem 2004; 5 (1): 68-75.

[22] Yin B, Ma H, Wang S, Chen S. Electrochemical synthesis of silver nanoparticles under protection of poly(n-vinylpyrrolidone). J Phys Chem B 2003; 107 (34): 8898-9004.

[23] Cheng JQ, Yao SW. Synthesis and characterization of silver nanoparticles by sonoelectrode-position. Rare Met 2005; 24 (4): 376-80.

[24] Socol Y, Abramson O, Gedanken A, Meshorer Y, Berenstein L, Zaban A. suspensive electrode formation in pulsed sonoelectrochemical synthesis of silver nanoparticles, Langmuir 2002;18 (12): 4736-40.

[25] Shchukin DG, Radtchenko IL. Sukhorukov G.B. Photoinduced reduction of silver inside microscale polyelectrolyte capsules. Chem Phys Chem 2003; 4 (10): 1101-03.

[26] Zhang LZ, Yu JC, Yip HY, *et al.* Ambient light reduction strategy to synthesize silver nanoparticles and silver-coated TiO_2 Langmuir 2003; 19 (24): 10372-80.

[27] Junior AM, de Oliveira HPM, Gehlen MH. Preparation of silver nanoprisms using poly(N-vinyl-2-pyrrolidone) as a colloid-stabilizing agent and the effect of silver nanoparticles on the photophysical properties of cationic dyes. Photochem Photobiol Sci 2003; 2 (9): 921-5.

[28] Jin RC, Cao YC, Hao EC, Metraux GS, Schatz GC, Mirkin. CA. Controlling anisotropic nanoparticle growth through plasmon excitation. Nature 2003; (425): 487-90.

[29] Mallick K, Witcomb MJ, Scurrell MS. Polymer stabilized silver nanoparticles: A photochemical synthesis route. J Mater Sci 2004; 39 (14): 4459-63.

[30] Kryukov AI, Zinchuk NN, Korzhak AV, Kuchmii SY. The effect of the conditions of catalytic synthesis of nanoparticles of metallic silver on their plasmon resonance. Theor Exp Chem 2003; 39 (1): 9-14.

[31] Cozzoli PD, Comparelli R, Fanizza E, Curri ML, Agostiano A, Laub D. Photocatalytic synthesis of silver nanoparticles stabilized by tio_2nanorods: A semiconductor/metal nanocomposite in homogeneous nonpolar solution. J Am Chem Soc 2004; 126 (12): 3868-79.

[32] Liu FK, Huang PW, Chang YC, Ko FH, Chu TC. Microwave-assisted synthesis of silver nanorods. J Mater Res 2004; 19 (2): 469-73.

[33] Yamamoto T, Yin HB, Wada Y, *et al.* Morphology-control in microwave-assisted synthesis of silver particles in aqueous solutions. Bull Chem Soc Jpn 2004; 77 (4): 757-61.

[34] Komarneni S, Li DS, Newalkar B, Katsuki H, Bhalla AS. Microwave-polyol process for Pt and Ag nanoparticles. Langmuir 2002; 18 (15): 5959-62.

[35] Yin H, Yamamoto T, Wada Y, Yanagida S. Large-scale and size-controlled synthesis of silver nanoparticles under microwave irradiation. Mater Chem Phys 2004; 83 (1): 66-70.

[36] Qin A, Jiang Z., Liu Q, Liao L, Jiang Y, Preparation of silver nanoparticles in the presence of polyacrylamide by microwave high-pressure synthesis method and its spectral properties. Chin J Anal Chem 2002; 30 (10): 1254-56.

[37] Hornebecq V, Antonietti M, Cardinal T, Treguer-Delapierre M. Stable silver nanoparticles immobilized in mesoporous silica. Chem Mater 2003; 15 (10): 1993-99.

[38] Choi SH, Lee SH, Hwang YM, Lee KP, Kang HD. Interaction between the surface of the silver nanoparticles prepared by γ-irradiation and organic molecules containing thiol group. Radiat Phys Chem 2003; 67 (1-3): 517-21.

[39] Li T, Park HG, Choi S-Ho. γ-Irradiation-induced preparation of Ag and Au nanoparticles and their characterizations. Mater. Chem. Phys. 2007; 105 (2-3): 325-30.

[40] Siskova K, Vlkova B, Turpin P-Y, Thorels A, Prochazka M. Laser Ablation of silver in aqueous solutions of organic species: probing ag nanoparticle-adsorbate systems evolution by surface-enhanced raman and surface plasmon extinction spectra. J Phys Chem C 2011; 115 (13): 5404-12

[41]   Zheng X, Zhu L, Wang X, Yan A, Xie Y. Simple mixed surfactant route for the preparation of noble metal dendrites. J Cryst Growth 2004; 260 (1-2): 255-62.

[42]   Al-Thabaiti SA, Al-Nowaiser FM, Obaid AY, Al-Youbi AO, Khan Z. Formation and characterization of surfactant stabilized silver nanoparticles: a kinetic study. Colloids Surf B Biointerfaces 2008; 67(2): 230-7

[43]   Zhang J, Han B, Liu M, et al. ltrasonication-induced formation of silver nanofibers in reverse micelles and small-angle X-ray scattering studies. J Phys Chem B 2003; 107 (16): 3679-83.

[44]   Maillard M, Giorgio S, Pileni MP. Tuning the size of silver nanodisks with similar aspect ratios: Synthesis and optical properties J Phys Chem B 2003; 107 (11): 2466-70.

[45]   Maillard M, Giorgio S, Pileni MP. Silver Nanodisks. Adv Mater 2002; 14 (15): 1084-86.

[46]   McLeod MC, McHenry RS, Beckman EJ, Roberts CB. Synthesis and stabilization of silver metallic nanoparticles and premetallic intermediates in perfluoropolyether/Co$_2$ reverse micelle systems. J Phys Chem B 2003; 107 (12): 2693-700.

[47]   Zhang W, Qiao X, Chen J. Synthesis of silver nanoparticles-effects of concerned parameters in water/oil microemulsion. Mater Sci Engin B 2007; 142 (1): 1-15

[48]   Kind C, Popescu R, Muller E, Gerthsen D, Feldmann C. Microemulsion-based synthesis of nanoscaled silver hollow spheres and direct comparison with massive particles of similar size Nanoscale 2010; 2 (10): 2223-29.

[49]   Gardea-Torresdey JL, Gomez E, Peralta-Videa JR, Parsons JG, Troiani H, Jose- Yacaman M. Alfalfa sprouts: A natural source for the synthesis of silver nanoparticles. Langmuir 2003; 19 (4): 1357-61.

[50]   Shankar SS, Ahmad A, Sastry M. Geranium leaf assisted biosynthesis of silver nanoparticles. Biotechnol Prog 2003; 19 (6): 1627-31.

[51]   Kowshik M, Ashtaputre S, Kharrazi S, et al. Extracellular synthesis of silver nanoparticles by a silver-tolerant yeast strain MKY3, Nanotechnology 2003; 14 (1): 95-100.

[52]   Ahmad A, Mukherjee P, Senapati S, et al. Extracellular biosynthesis of silver nanoparticles using the fungus fusariumoxy-sporum. Colloids Surf B Biointerfaces 2003; 28 (4): 313-8.

[53]   Bhainsa KC, DSouza SF. Extracellular biosynthesis of silver nanoparticles using the fungus Aspergillus fumigates. Colloids Surf B Biointerfaces 2006; 47 (2): 160-4.

[54]   Shankar SS, Rai A, Ahmad A, Sastry M, Rapid synthesis of Au, Ag, and bimetallic Au core-Ag shell nanoparticles using Neem (Azadirachta indica) leaf broth. J Colloid Interface Sci 2004; 275 (2): 496-502.

[55]   Jang H D. Experimental study of synthesis of silica nanoparticles by a bench-scale diffusion flame reactor. Powder Technol 2001; 119 (2-3): 102-8.

[56]   Jiu J, Isoda S, Adachi M, Wang F, Preparation of TiO$_2$ nanocrystalline with 3- 5 nm and application for dye-sensitized solar cell. J Photochem Photobiol A Chem 2007; 189 (2-3): 314-21.

[57]   Corradi AB, Bondioli F, Ferrari AM' Focher B, Leonelli C, Synthesis of silica nanoparticles in a continuous-flow microwave reactor. Powder Technol 2006; 167 (1): 45-8.

[58]   Guo J, Liu X, Cheng Y, Li Y, Xu G, Cui P. Size-controllable synthesis of monodispersed colloidal silica nanoparticles via hydrolysis of elemental silicon. J Colloid Interface Sci 2008; 326 (1): 138-42.

[59]   Medintz IL, Uyeda HT, Goldman ER, Mattoussi H. Quantum dot bioconjugates for imaging, labeling and sensing. Nat Mater 2005; (4): 435-46.

[60]   Caruthers SD, Wickline SA, Lanza GM. Nanotechnological applications in medicine. Curr Opin Biotech 2007; 18 (1): 26-30.

[61]   Marquis BJ, Love SA, Braun KL, Haynes CL. Analytical methods to assess nanoparticle toxicity. Analyst 2009; 34 (3); 425-39

[62]   Borm PJA, Robbins D, Haubold S, et al. The potential risks of nanomaterials: a review carried out for ECETOC. Part Fibre Toxicol 2006; (3): 11-35.

[63]   Hansen SF, Michelson ES, Kamper A, Borling P, Stuer-Lauridsen F, Baun A. Categorization framework to aid exposure assessment of nanomaterials in consumer products. Ecotoxicology 2008; 17 (5): 438-47.

[64]   Hristozov D, Malsch I. Hazards and risks of engineered nanoparticles for the environment and human health. Sustainability 2009; (1): 1161-94.

[65]   Maynard AD, Aitken RJ, Butz T, et al. Safe handling of nanotechnology. Nature 2006; (444): 267-9.

[66]   Tsuji JS, Maynard AD, Howard PC, et al. Research strategies for safety evaluation of nanomaterials, part IV: Risk assessment of nanoparticles. Toxicol Sci 2006; 89 (1): 42-50.

[67]   Raviraja NS, Kandikere RS. Nanotoxicity: Threat posed by nanoparticles. Curr Sci 2007; 93(6): 769-70.

[68]   John Garrod. Characterizing the potential risks posed by engineered nanoparticles: A second UK government research report; Department for Environment, Food and Rural Affairs Nobel House, 17 Smith Square, London SW1P 3JR December 2007.

[69] Colvin VL. The potential environmental impact of engineered nanomaterials. Nat Biotechnol 2003; 21(10): 1166-70.

[70] Nanotec.org [homepage on the internet].UK; Royal Academy of Engineering [updated: 5 January 200; cited: 02 June 2001]: Final report; nanoscience and nanotechnologies: opportunities and uncertainties. Available from http://www.nanotec.org.uk/finalReport.htm

[71] Krätschmer W, Lamb LD, Fostiropoulos K, Huffman DR. Solid C_{60}: A new form of carbon. Nature 1990; (347): 354-8.

[72] Howard JB, McKinnon JT, Makarovsky Y, *et al.* Groundwork for a rational synthesis of C_{60}: Cyclodehydrogenation of a $C_{60}H_{30}$ polyarene. Science 2001; 294 (5543): 828-31.

[73] Scott LT, Boorum MM, McMahon BJ, *et al.* A rational chemical synthesis of C_{60}. Science 2002; 295 (5559): 1500-03.

[74] Scott, LT. Methods for the chemical synthesis of fullerenes. Angew Chem Int Ed 2004; 43 (38): 4994-5007.

[75] Iijima, S. Helical microtubules of graphitic carbon. Nature 1991; (354): 56-8.

[76] Burgos J C, Reyna H, Yakobson B I, Balbuena P B. Interplay of Catalyst Size and Metal-Carbon Interactions on the Growth of Single-Walled Carbon Nanotubes. J Phys Chem C 2010;114 (15): 6952-58

[77] Oncel C, Yurum Y. Carbon Nanotube Synthesis *via* the Catalytic CVD Method: A Review on the Effect of Reaction Parameters. Fuller. Nanotub. Carbon Nanostruct 2006; 14 (1): 17-37.

[78] Ivanov V, Nagy JB, Lambin PH, *et al.* The study of carbon nanotubules produced by catalytic method. Chem Phys Lett 1994; 223(4): 329-35.

[79] Faraji AH, Wipf P, Nanoparticles in cellular drug delivery. Bioorg Med Chem 2009; 17(8): 2950-60.

[80] Sen R, Govindaraj A, Rao CNR. Carbon nanotubes by the metallocene route. Chem Phys Lett 1997; 267(3-4): 276-80.

[81] Jang J, Yoon H. Fabrication of magnetic carbon nanotubes using a metal- impregnated polymer precursor. Adv Mater 2003; 15(24): 2088-91.

[82] Satishkumar BC, Govindaraj A, Sen R, Rao CNR. Single-walled nanotubes by the pyrolysis of acetylene-organometallic mixtures. Chem Phys Lett 1998; 293(1-2): 47-52.

[83] Bccresearch.com [homepage on the internet]. Carbon Nanotubes: Technologies and Global Market. Massachusett; BCC Research market forecasting [updated 14 January 2011; cited: 02 June2011]. Available from http://www.bccresearch.com/editors/RGB-290.html

[84] Bruchez MJ, Moronne M, Gin P, Weiss S, Alivisatos AP. Semiconductor nanocrystals as fluorescent biological labels. Science 1998; 281(5385): 2013-16.

[85] Sau T K, Pal A, Jana NR, Wang ZL, Pal T. Size controlled synthesis of gold nanoparticles using photochemically prepared seed particles. J Nanopart Res 2001; (3): 257-61.

[86] Ullman A. Formation and structure of self-assembled monolayers. Chem Rev 1996; 96 (4):1533-54.

[87] Tanga T, Krysmanna MJ, Hamley IW. *In situ* formation of gold nanoparticles with a thermoresponsive block copolymer corona. Colloid Surf A-Physicochem Eng Asp 2008; 317 (1-3): 764-7.

[88] Moyer CA. A treatment of burns. Trans Stud Coll Physicians Philadelphia 1965; 33(2): 53-103.

[89] Wang WN, Lenggoro IW, Terashi Y, Kim TO, Okuyama K. One-step synthesis of titanium oxide nanoparticles by spray pyrolysis of organic precursors. Mater Sci Engin B 2005; 123 (3): 194-202.

[90] Kim KD, Kim HT. Synthesis of titanium dioxide nanoparticles using a continuous reaction method. Colloid Surf. A-Physicochem Eng Asp 2002; 207 (1-3): 263-9.

[91] Mahshid S, Askari M Ghamsari MS. Synthesis of TiO_2 nanoparticles by hydrolysis and peptization of titanium isopropoxide solution. J Mater Process Technol 2007; 189 (1-3): 296-100.

[92] Dreesen L, Colomer JF, Limage H, Giguère A, Lucas S. Synthesis of titanium dioxide nanoparticles by reactive DC magnetron sputtering, Thin Solid Films 2009; 518 (1): 112-5.

[93] Figgemeier E, Kylberg W, Constable E, *et al.* Titanium dioxide nanoparticles prepared by laser pyrolysis: Synthesis and photocatalytic properties. Appl Surf Sci 2007; 254 (4): 1037-41.

[94] Prasad K, Pinjari DV, Pandit AB, Mhaske ST, Synthesis of titanium dioxide by ultrasound assisted sol-gel technique: Effect of amplitude (power density) variation. Ultrason Sonochem 2010; 17 (4): 697-703.

[95] Jang HD. Experimental study of synthesis of silica nanoparticles by a bench-scale diffusion flame reactor. Powder Technol 2001; 119 (2-3): 102-8.

[96] Corradi AB, Bondioli F, Ferrari AM Focher B, Leonelli C. Synthesis of silica nanoparticles in a continuous-flow microwave reactor. Powder Technol 2006; 167 (1): 45-8.

[97] Rao KS, El-Hami K, Kodaki T, Matsushige K, Makino KA novel method for synthesis of silica nanoparticles. J Colloid Interface Sci 2005; 289 (1): 125-31.

[98]   Chang H, Park J-H, Jang HD, Flame synthesis of silica nanoparticles by adopting two-fluid nozzle spray, Colloid Surf. A-Physicochem Eng Asp 2008; (313-314): 140-4.

[99]   Gupta AK, Gupta M. Synthesis and surface engineering of iron oxide nanoparticles for biomedical applications. Biomaterial 2005; 26 (18): 3995-21

[100] Guin D, Manorama SV. Room temperature synthesis of monodispersed iron oxide nanoparticles. Mater Lett 2008; 62 (17-18): 3139-42.

[101] Iijima M, Sato K, Kurashima K, Ishigaki T, Kamiya H. Low-temperature synthesis of redispersible iron oxide nanoparticles under atmospheric pressure and ultradense reagent concentration. Powder Technol 2008; 18 (1): 45-50.

[102] Figuerola A, Corato RD, Manna L, Pellegrino T. From iron oxide nanoparticles towards advanced iron-based inorganic materials designed for biomedical applications, Pharma Res 2010; 62 (2): 126-43.

[103] Wiesner MR, Bettero J-Y. Environmental nanotechnology; application and impacts of nanomaterials, McGrew Hill professional publishing companies, 2007.pp 446-47.

[104] Aillon KL, Xie Y, El-Gendy N, Berkland CJ., Forrest ML. Effects of nanomaterial physicochemical properties on *in vivo*toxicity. Adv Drug Deliv Rev 2009; 61 (6): 457-66.

[105] Alivisatos P. The use of nanocrystals in biological detection. Nat Biotechnol 2004; (22): 47-52.

[106] Hardman R, A toxicologic review of quantum dots: Toxicity depends on physicochemical and environmental factors. Environ Health Perspect 2006; 114 (2): 165-72.

[107] Lanone S, Boczkowski J. Biomedical applications and potential health risks of nanomaterials: molecular mechanisms. Curr Mol Med 2006; 6 (6): 651-63.

[108] Powers KW, Brown SC, Krishna VB, Wasdo SC, Moudgil BM, Roberts SM, Research strategies for safety evaluation of nanomaterials. part VI. Characterization of nanoscale particles for toxicological evaluation, Toxicol. Sci. 2006; 90 (2): 296-303.

[109] Garnett MC, Kallinteri P, Nanomedicines and nanotoxicity: Some physiological principles, Occup Med 2006; 56 (5): 307-11.

[110] Fischer HC, Chan WCW. Nanotoxicity: The growing need for in-vivo study, Curr Opin Biotechnol 2007; 18 (6): 565-71.

[111] Tian F, Cui D, Schwarz H, Estrada GG, Kobayashi H. Cytotoxicity of single-wall carbon nanotubes on human fibroblasts. Toxicol *In vitro* 2006; 20 (7): 1202-12.

[112] Soto K, Garza KM, Murr LE, Cytotoxic effects of aggregated nanomaterials, Acta Biomater 2007; 3 (3): 351-8.

[113] Jones CF, Grainger DW. *In vitro* assessments of nanomaterial toxicity, Adv Drug Deliv Rev 2009; 61 (6): 438-56.

[114] Davoren M, Herzog E, Casey A, *et al. In vitro* toxicity evaluation of single walled carbon nanotubes on human A549 lung cells. Toxicol *In vitro* 2007; 21 (3): 438-48.

[115] Wilson MR, Lightbody JH, Donaldson K, Sales J, Stone V. Interactions between ultrafine particles and transition metals *in vivo* and *in vitro*. Toxicol Appl Pharmacol 2002; 184 (3):172-9.

[116] Geiser M, Rothen-Rutishauser B, Kapp N, *et al.* Ultrafine particles cross cellular membranes by nonphagocytic mechanisms in lungs and in cultured cells. Environ Health Perspect 2005; 113 (11): 1555-60.

[117] Soto KF, Carrasco A, Powell TG, Murr LE, Garza KM. Biological effects of nanoparticulate materials. Mater Sci Eng C 2006; 26 (8): 1421-27.

[118] Bogner A, Thollet G, Basset D, Jouneau PH, Gauthier C. Wet STEM: A newdevelopment in environmental SEM for imaging nano-objects included in a liquid phase. Ultramicroscopy 2005; 104 (3-4): 290-301.

[119] Haider S, Saeed K, Farmer BL. Swelling and electroresponsive characteristics of gelatin immobilized onto multi-walled carbon nanotubes. Sens Actuators B 2007; 124 (2): 517-28.

[120] Haider S, Park SY. Preparation of the electrospun chitosan nanofibers and their applications to the adsorption of Cu (II) and Pb (II) ions from an aqueous solution. J Membr Sci 2009; 328 (1-2): 90-6.

[121] Saeed K, Park SY, Lee HJ, Baek JB, Huh WS. Preparation of electrospun nanofibers of carbon nanotube/polycaprolactone nanocomposite. Polymer 2006; 47 (23): 8019-25.

[122] Alothman ZA, Bukhari N, Haider S, Wabaidur SM, Alwarthan AA.Spectrofluorimetric determination of fexofenadine hydrochloride in pharmaceutical preparation using silver nanoparticles. Arab J Chem 2010; 3 (4): 251-5.

[123] Omer M, Haider S, Park SY, A novel route for the preparation of thermally sensitive core-shell magnetic nanoparticles. Polymer 2011; 52 (1): 91-7.

[124] El-Sayed MA. Small is different: shape-, size-, and composition-dependent properties of some colloidal semiconductor nanocrystals. Acc. Chem. Res. 2004; 37 (5): 326-33.

[125] Lin W, Huang Y, Zhou XD, Ma Y. *In vitro* toxicity of silica nanoparticles in human lung cancer cells. Toxicol Appl Pharmacol 2006; 217 (3): 252-9.

[126]   Castner D.G. Surface functionalization and characterization of nanoparticles for biomedical applications. National ESCA & Surface Analysis Center for Biomedical Problems, Departments of Bioengineering and Chemical Engineering, University of Washington 2008.

[127]   Wiesner MR, Lowry GV, Alvarez P, Dionysiou D, Biswas P. Assessing the risks of manufactured nanomaterials. Environ Sci Technol. 2006; 40 (14): 4337- 45.

CHAPTER 8

Toxicologic and Environmental Issues Related to Nanotechnology Development

Ibrahim Abdulwahid Arif[1], Haseeb Ahmad Khan[1,2], Salman Al Rokayan[2,3], Abdullah Saleh Alhomida[2], Mohammad Abdul Bakir[1] and Fatima Khanam[4]

[1]*Prince Sultan Research Chair for Environment and Wildlife, College of Science, King Saud University, Riyadh, Saudi Arabia;* [2]*Department of Biochemistry, College of Science, King Saud University, Saudi Arabia;* [3]*King Abdullah Institute for Nanotechnology, King Saud University, Riyadh, Saudi Arabia and* [4]*Department of Chemistry, College of Science, King Saud University, Riyadh, Saudi Arabia*

Abstract: This decade has seen revolutionary developments in the field of nanotechnology with newer and diverse applications of nanoparticles appearing everyday. Novel nanomaterials are emerging with different characteristics and compositions for specific applications such as cosmetics, drug delivery, imaging, electronic *etc.* However, little attention is being paid to understand, assess and manage the environmental impact of nanoparticles. Currently the information about toxicity of nanoparticles and their environmental fate in air, water and soil is severely lacking. Inhalation, ingestion and dermal penetration are the potential exposure routes for nanoparticles whereas particle size, shape, surface area and surface chemistry collectively define the toxicity of nanoparticles. Several studies have shown excessive generation of reactive oxygen species as well as transient or persistent inflammation following exposure to various classes of nanoparticles. Increased production and intentional (sunscreens, drug-delivery) or unintentional (environmental, occupational) exposure to nanoparticles is likely to increase the possibilities of adverse health effects. The major environmental concerns include exposure assessment, biological fate, toxicity, persistence and transformation of nanoparticles. Thus, the novel nanomaterials need to be biologically characterized for their health hazards to ensure risk-free and sustainable implementation of nanotechnology.

Keywords: Nanotechnology, Nanoparticles, Toxicity, Environmental impact, Safety, Adverse effects, Health hazards, Biological characterization.

NANOTECHNOLOGY: AN ANCIENT BACKGROUND OF RECENT TECHNOLOGY

Although nanotechnology is a fairly new field, nanomaterials are not. Gold (Au) and silver (Ag) nanoparticles (NPs) had been used in Persia in the 10th century BC to fabricate ceramic glazes to provide a lustrous or iridescent effect. This technique was then brought to Spain, where it was improved by the Moors during the 14th century, before finally spreading throughout much of Europe. In addition, over 5000 years ago, the Egyptians ingested Au NPs for mental and bodily purification. In the present time, the NPs are manufactured for need-based applications and so emerged the term, engineered nanoparticles (ENPs). In general, NPs are usually defined by their core materials such as organic and inorganic. Organic NPs can be further defined as fullerenes (C60 and C70 and derivatives) and carbon nanotubes (multi-walled or single-walled CNTs), while inorganic NPs can be divided into metal oxides (of iron, zinc, titanium, *etc.*), metals (silver, gold, *etc.*) and quantum dots (cadmium sulfide, cadmium selenide, *etc.*) [1]. Other classifications and terminologies are also used to refer some specific groups of nanomaterials *i.e.,* nanocrystals (single crystal nanoparticles) and different morphologies such as spheres, pyramids and cubes. Some of these NPs, such as the ones based on metals and fullerenes, offer the possibility of manipulating their surfaces in order to introduce specific functionalities for further applications [2].

MERITS AND DEMERITS OF NANOTECHNOLOGY

With a dormant ancient history, nanotechnology is a major innovative scientific and economic growth area

---

*****Address correspondence to Haseeb Ahmad Khan:** Department of Biochemistry, College of Science, King Saud University, P.O. Box 2455, Riyadh 11451, Saudi Arabia; Tel: +966-1-4675859; E-mail: khan_haseeb@yahoo.com of recent time. Although nano-sized particles have always occurred in nature, the latest developments are in the production and use of ENPs which are finding applications in a wide range of areas including cosmetics, medicine, food and food packaging, bioremediation, paints, coatings, electronics and fuel catalysts and water treatment [3, 4]. The drugs encapsulated into nanoparticles results in a clump-free, stable and water-soluble material due to a very large surface to volume ratio [5, 6]. Nanoparticle-mediated drug delivery systems are being developed for preventive treatment of the oxidative damage implicated in various neurodegenerative diseases such as Alzheimer's disease, Parkinson's disease and Wilson's disease [7, 8]. Lecaroz *et al.* [9] have suggested the benefits of nanocarriers loaded with gentamicin for treatment of intracellular pathogens such as brucellosis. Pertuit *et al.* [10] have shown the therapeutic potential of 5-amino salicylic acid bound nanoparticles for the treatment of inflammatory bowel disease while NP formulations allowed to significantly reduce the dose of active agent. Nanogold particles have exerted antiangiogenic activities and subsequently reduced macrophage infiltration and inflammation, which resulted in attenuation of arthritis in a rat model [11]. Novel nanomaterials are also being explored for potential therapeutic and diagnostic applications in cancer treatment and diagnosis where the nanoscale properties facilitate entry and intracellular transport to specific target sites [5]. Nanotechnologies hold great promise for reducing the production of wastes and industrial contamination and improving the efficiency of energy production and use [1].

The potential for nanotechnology is believed to be practically limitless and the potential for profiting from creating and marketing these advances is the driving force behind an incredibly rapid rush to deliver these applications to the marketplace [12]. However, the production, use and disposal of manufactured NPs lead to discharges into air, soils and aquatic systems. Therefore, it is crucial to investigate their transport into and through the environment and their impacts on environmental health [1]. The indiscriminate use of ENPs with unknown toxicological properties might pose a variety of hazards for environment, wildlife and human health. Our knowledge of the harmful effects of nanoparticles is still very limited and at present no specific regulations have been developed for ENPs usage. The main concern is whether unknown risks of ENPs, in particular their impact on health and the environment outweigh their potential benefits for society [13]. Since the nanomedicine and nanotoxicology are two sides of the same coin [14], the worth of this coin depends on its prudent use.

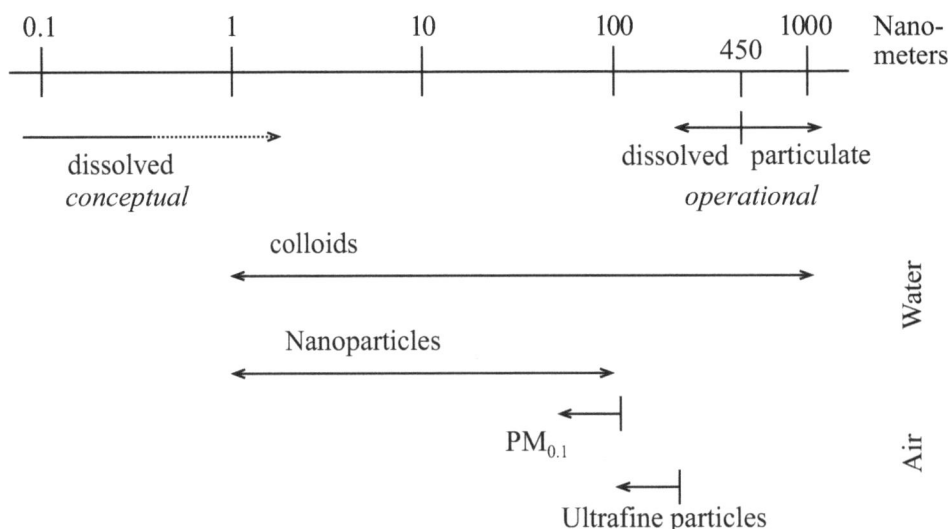

Figure 1: Size distribution of various particles.

UNIQUE PROPERTIES OF ENGINEERED NANOPARTICLES

Engineered nanoparticles are usually defined as manufactured particles with approximate dimension between 1 and 100nm (Fig. **1**) [15, 16]. The number of atoms at the surface and the physical properties of NPs differ from larger materials [17]. Properties associated with the bulk materials are averaged properties, such as density, resistivity and magnetisation and the dielectric constant however many properties of these materials change over at the NP scale [18]. Nanoparticles have a proportionately very large surface area and this surface can have a high affinity for metals and organic pollutants such as polycyclic aromatic hydrocarbons.

Nanomaterials also present different interesting morphologies such as spheres, tubes, rods and prisms. Nanotechnology includes the integration of these nanoscale structures into larger material components and systems, keeping the control and construction of new and improved materials at the nanoscale [19]. Among these novel nanomaterials, nanoparticles play an important role in nanotechnology advances [1].

RISK OF EXPOSURE TO NANOPARTICLES

There is potential risk for exposure of humans and the environment to nanoparticles throughout their life cycle, starting from manufacture to disposal. Accidental spillages or permitted release of industrial effluents in waterways and aquatic systems may result in direct exposure to nanoparticles of humans *via* skin contact, inhalation of water aerosols and direct ingestion of contaminated drinking water or particles adsorbed on vegetables or other foodstuffs [20, 21]. The potential sources of NPs exposure and their routes are outlined in Fig. **2**. There is a very large body of evidence that small particles produced by combustion processes known as "ultrafines", or nanoparticles by an older name, can be dangerous to human health [22-25].

Due to the relatively recent emergence of nanotechnology, government agencies such as the U.S. Bureau of Labor Statistics have not yet published reliable estimates of the number of workers currently involved in researching and manufacturing of nanomaterials. The nanotechnology related trade magazine Small Times estimates that there are currently approximately 25, 000 workers involved in companies that work strictly with nanotechnology [12]. In a National Advisory Committee on Occupational Safety and Health meeting in December 2004, it was reported that a Rand Corporation study estimated over 2, 000, 000 workers would be involved in nanotechnology related jobs worldwide within 10 to 15 years [21].

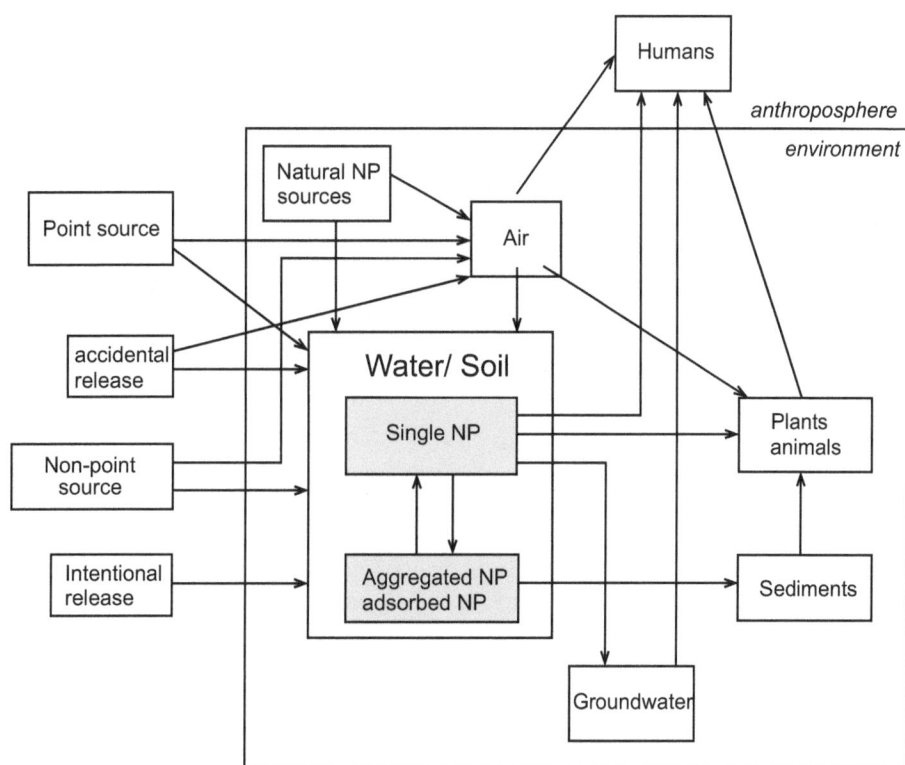

Figure 2: Nanoparticle pathways from anthroposphere into environment, reactions in the environment and exposure of humans. Reproduced with permission from Elsevier; Nowack and Bucheli [26].

The human populations potential at risk of exposure to nanoparticles include researchers and workers in nanomaterials industry, their family members, consumers and general public [12]. Risk management should be an integral part of an occupational safety and health program, which is based on recognition of the nanomaterial hazards, evaluation of the exposure potentials, and application of control measures to reduce the risk [27].

ROUTE OF ENTRY INTO LIVING ORGANISMS

Uptake of nanoparticles by inhalation is the major route of their entry into the body [12, 21, 28-30]. While inhalation is the primary exposure route of concern, certain applications and conditions raise the risk of dermal absorption and pose a potential exposure route into the body. Sunscreens containing titanium oxide nanoparticles have been found to penetrate the epidermal skin layer [31]. Once the particles have reached the dermal layer, it is likely that diffusion would allow for access to the bloodstream resulting in the same concerns for translocation encountered through inhalation exposure [12]. Although not the primary one, ingestion may also be considered as an important route of nanoparticles exposure, either by accidental cause or due to hand-to-mouth transfer in the workplace. However, in aquatic animals there may be other routes of entry such as direct passage across gills [32].

NPs may travel to other places and interact with tissues prolonging their residence in different body organs. They may also reach potentially sensitive target sites such as the heart and brain. Moreover, the distribution of NPs in the liver, kidney, and immune-modulating organs (spleen, bone marrow) has also been observed (Fig. **3**). Combustion-derived nanoparticles and their components have the potential to translocate to the brain and also the blood, and thereby reach other targets such as the cardiovascular system, spleen and liver [33]. Elder *et al.* [34] have concluded that the olfactory neuronal pathway is efficient for translocating inhaled Mn oxide as solid ultrafine particles to the central nervous system.

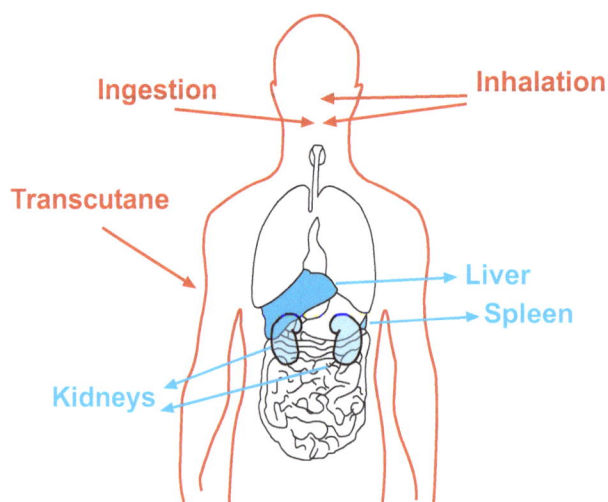

Figure 3: Biodistribution of nanoparticles. Reproduced with permission from Elsevier; Casals *et al.* [13].

Some insoluble nanoparticles can pass through the different protective barriers, distribute in the body and accumulate in several organs. At the cellular level, most internalisation of nanoparticles occurs *via* endocytosis [35-37]. The inherent ability of the particles to penetrate the body and cells provides the potential routes for the delivery of nanoparticle-bound toxic pollutants to sites where they would not reach in the normal unbound form [35-38].

PHYSICOCHEMICAL AND BIOCHEMICAL ASPECTS OF NANOTOXICITY

Size appears to effect particle uptake by cells and penetration across cell barriers. The surface area is related to the potential of particles to generate inflammation. The shape, especially fiber shape appears to promote inflammation, fibrosis and granuloma formation *in vivo*, linked to phagocytosis and pro-inflammatory signaling *in vitro*. Positively charged particles are often more toxic than negative particles, and some particles appear to leach soluble toxic components. Rothen-Rutishauser *et al.* [39] have provided evidence that the translocation and entering characteristics of particles are size-dependent whereas intracellular localization of nanoparticles depends on the particle material. The pulmonary toxicities of alpha-quartz particles appeared to correlate better with surface activity than particle size and surface area [40]. Gojova *et al.* [41] have demonstrated that inflammation in human aortic endothelial cells (HAECs) following acute exposure to metal oxide nanoparticles depends on particle composition. For instance, Fe_2O_3 nanoparticles fail to provoke an inflammatory response in HAECs however Y_2O_3 and ZnO nanoparticles elicit a pronounced inflammatory response above a threshold concentration and may be cytotoxic leading to considerable cell death [41]. Wagner *et al.* [42] have observed that aluminum metal NPs exhibit greater toxicity and more significantly diminish the phagocytotic ability of macrophages as compared to aluminum oxide NPs. The interaction of nanoparticles and pollutants for cellular uptake is shown in Fig. **4**.

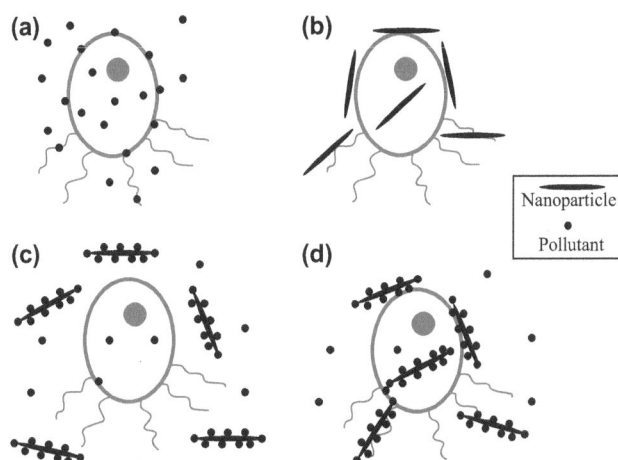

Figure 4: Scheme of the interactions of NP, pollutants, and organisms. (a) adsorption and uptake of pollutant, (b) adsorption and uptake of NP, (c) adsorption of pollutants onto NP and reduction in pollutant uptake by organisms and (d) adsorption of NP with adsorbed pollutant and possible uptake of pollutant-NP. Reproduced with permission from Elsevier; Nowack and Bucheli [26].

Nanofibres and nanotubes may present special problems with respect to their retention in cells and tissues, governed by a threshold length of the fibre that is critical for induction of adverse biological effects [43]. Although the size of the nanoparticles itself can be a prime factor in direct toxicity and pathology, its biodegradability may be a further significant factor in governing harmful biological effects [21, 43, 44]. The manufactured nanoparticles may present living systems with a uniquely novel challenge, as these materials were not generally encountered by living organisms during the course of biological evolution [21, 28-30]. So, there has been little or no selection pressure for defensive or protective systems to counter any adverse properties that such particles may present beyond those already presented by naturally occurring combustion products, volcanic ash, toxic metals and organic xenobiotics [45]. Moreover, the toxicity of some nanoparticles may be related to the surface reactivity of the particles, in influencing the development of inflammatory and cytotoxic responses [40, 46].

Nanoparticles situated in the endoplasmic reticulum, Golgi and lysosomal systems could conceivably act as foci of oxidative damage that could not readily be expelled from the cell; while generation of free radicals could lead to cellular damage and ultimately organelle dysfunction. Following the inhalation of NPs, a number of diverse processes involving physical translocation of entered particles, some of them showing significant particle-size dependence, have been observed. The most prevalent mechanism for clearance of NPs in the alveolar region is mediated by macrophages, through phagocytosis of deposited particles. Essentially, all the particles are phagocytized by alveolar macrophages 6-12 h after deposition, and subsequently cleared by the slow alveolar clearance mechanism [47]. The retention half-time of solid particles in the alveolar region based on this mechanism is about 2 months in rats and up to 2 years in humans [48]. Another possible mechanism involves transcytosis across epithelia of the respiratory tract into the interstitium and access to the blood circulation directly or *via* the lymphatic system, resulting in distribution throughout the body [48]. Moreover, nanoparticles may penetrate the red blood cell membrane by a still unknown mechanism different from phagocytosis and endocytosis [49].

The translocation of by the blood circulation could be responsible for the deleterious effects on the cardiovascular system induced by inhaled ambient ultrafine particles. However, Brown *et al.* [50] have suggested that the presence to foreign particle in heart may not be the condition for cardiotoxicity as the particle-induced inflammation in the lungs my indirectly provoke cardiac toxicity. Ultrafine particles of two very different materials induced inflammation and epithelial damage to a greater extent than their fine counterparts; the effect of ultrafine carbon black was greater than ultrafine titanium dioxide, suggesting that there are differences in the likely harmfulness of different types of ultrafine particle [51]. The epithelial injury and toxicity after exposure to ultrafine particles were associated with the development of inflammation [51]. The plasma clearance, distribution, and metabolic profiles of lysine dendrimers are significantly influenced by the structure and charge of the capping groups, particularly larger arylsulfonate-capped lysine dendrimers are rapidly opsonized and initially cleared from the plasma by the reticuloendothelial organs [52].

TOXICITY AND HEALTH HAZARDS OF NANOPARTICLES

The small size of ENPs facilitates their uptake into cells and translocation across epithelial and endothelial cells. ENPs may travel to other places in body and interact with tissues prolonging their stay in the body. Those particles could thus reach potentially sensitive target sites including heart and brain. The US Environmental Protection Agency has attributed 60, 000 deaths per year to the inhalation of atmospheric nanoparticles [25, 53]. Studies have found the accumulation of ENPs in organs with high phagocytic activity, mainly in liver (up to 90%), kidney and spleen. Toxic effects have been documented at the pulmonary, cardiac, reproductive, renal, cutaneous and cellular levels. Significant accumulations have been shown in the lungs, brain, liver, spleen and bones. A precautionary approach is required with individual evaluation of new nanomaterials for risk to the health of the environment [45]. Although current toxicity testing protocols may be applicable to identify harmful effects associated with nanomaterials, research into new methods is necessary to address the special properties of nanomaterials. The results of the rodent studies collectively showed that regardless of the process by which carbon nanotubes were synthesized and the types and amounts of metals they contained, they were capable of producing inflammation, epithelioid granulomas, fibrosis, biochemical and toxicological changes in the lungs [54, 55].

Nanoparticles have a proportionately very large surface area that can have a high affinity for inorganic and organic toxicants such as iron and polycyclic aromatic hydrocarbons, respectively [56]. Jia *et al.* [57] have shown that carbon nanomaterials with different geometric structures exhibit quite different cytotoxicity and bioactivity *in vitro*, although they may not be accurately reflected in the comparative toxicity *in vivo*. Inoue *et al.* [58] have suggested that that acute exposure to diesel nanoparticles exacerbates lung inflammation induced by lipopolysaccharides. Reeves *et al.* [59] have revealed that the toxic effects of nanoparticulate TiO_2 are most likely due to hydroxyl radical formation; UVA irradiation of TiO_2-treated cells causes further increases in DNA damage. Instillation of a small dose of ultrafine colloidal silica particles causes transient acute moderate lung inflammation and tissue damage due to the mechanisms related to oxidative stress and apoptosis [60]. While increased TNF-alpha production was seen to accompany ambient ultrafine particle-induced oxidant injury, cationic polystyrene nanospheres induced mitochondrial damage and cell death without inflammation [61].

Nanoparticles can also increase the toxicity of other materials, for instance zinc chloride-induced stimulation of TNF-alpha is synergistically enhanced by 14 nm carbon black [62]. Waldman *et al.* [63] have shown that 100 nm carbon particles with nonextractable surface-bound iron are far more biologically active, toxic and proinflammatory than unbound particles. Sayes *et al.* [64] failed to observe a satisfactory correlation between in-vitro and in-vivo toxicology data suggesting that *in vitro* cellular systems will need to be further developed, standardized, and validated relative to *in vivo* effects in order to provide useful screening data on the relative toxicity of inhaled particle types. For any application and future developments, a key issue is therefore accurate assessment of the potential toxicity of NPs [13].

ECOTOXICITY AND ENVIRONMENTAL FATE OF NANOMATERIALS

Nanomaterials are prone to enter the environment for several reasons. The mass production of nanomaterials in the recent years paves the path of their likely entry into the environment from manufacturing effluent or from spillage during shipping and handling [48]. The use of nanoparticles in personal-care products such as cosmetics and sunscreens allow them to enter the environment on a continual basis from washing off of consumer products [65]. They are being used in electronics, tires, fuel cells, and many other products, and it is unknown whether some of these materials may leak out or be worn off over the period of use. Moreover, their application in disposable materials such as filters and electronics increases the risk of their environment entry through landfills and other methods of disposal [48].

The environmental fate and ecotoxicity of ENPs is influenced by a number of properties, including particles size and size distribution, solubility and state of aggregation, elemental composition, mass and number concentrations, shape and crystal structure, surface area, charge and chemistry, presence of impurities and environmental conditions (*e.g.,* pH, temperature, light). As a result, ENPs of the same material but with different crystal structure, surface coating, shape or size can have very different behavior, uptake, effects and environmental impact. The higher level consequences for damage to animal health, ecological risk and possible food chain risks for humans may be anticipated based on known behaviors and toxicities for inhaled and ingested nanoparticles.

The dispersion of NPs in the environment strongly depends on their ability to remain independent avoiding sedimentation, agglomeration or disintegration (Fig. **5**). Inorganic NPs are not very common in nature due to their instability, thus their fate is mainly to aggregate with other materials, change characteristics [66] or disintegrate into atomic and molecular species [67], which in turn transformed into stable species or incorporated into other materials. As the aggregate or disintegrated species can be toxic in some cases, NPs may also be considered as a pro-toxin rather than a toxin itself [13]. Auger *et al.* [68] have shown that diesel exhaust particles are more potent than PM2.5 for producing reactive oxygen species (ROS) while the airway epithelial cells exposed to these particles augment the local inflammatory response in the lung but cannot alone initiate a systemic inflammatory response.

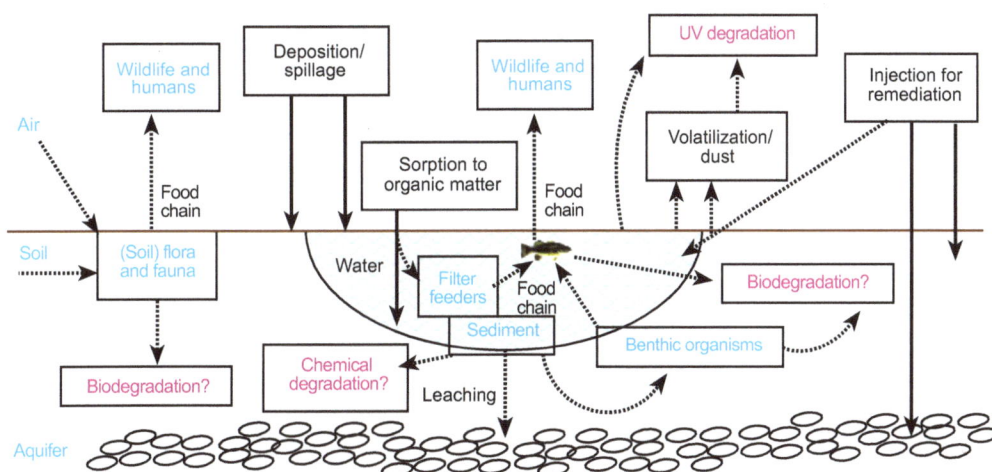

Figure 5: Routes of exposure, uptake, distribution, and degradation of nanoparticles in the environment. Reproduced with permission from EHP; Oberdörster *et al.* [48].

Although nanoparticles have received much attention for their applications in biological studies, various studies have shown the potential adverse effects of nanoparticles on human health and the environment

[69]. It is likely that waste generated during the production and use of these nanoparticles will appear eventually in various environments [70]. A recent report of the Royal Society and the Royal Academy of Engineering on nanoscience and nanotechnology [71] is drawing the attention of environmental managers and policy makers on the possible toxicological and pathological risks to human health and to the environment presented by novel nanomaterials resulting from nanotechnologies or from the interface between biotechnology and nanotechnology [21, 28, 29, 71]. As the nanotechnology industries propagate with large production rates, the entry of nanoscale products and by-products into the aquatic environment would be inevitable. Moore [45] has anticipated that a major challenge for ecotoxicologists will be the derivation of toxicity thresholds for nanomaterials and determining whether or not currently available biomarkers of harmful effect will also be effective for environmental nanotoxicity. However, the interaction of NP with toxic, organic compounds can both amplify as well as alleviate the toxicity of the compounds. Thus, despite harmful effects, ENPs can also have an advantageous role in the environment. Our knowledge in this area is still limited, thus more coordinated research is required to gain a better understanding of the factors and processes affecting ENP fate and effects in the environment as well as to develop more usable, robust and sensitive methods for characterization and detection of ENPs in environmental systems [72].

CONCLUSION

Although there is a general recognition that nanotechnology has the potential to advance science, quality of life and to generate substantial financial gains, a number of reports have also suggested that potential toxicity should be considered as a primary concern for safe use of nanoparticles. The current knowledge of the toxic effects of nanoparticles is relatively limited. A precautionary approach is therefore required with individual evaluation of new nanomaterials for risk to the environment. Although current toxicity testing protocols may be applied to identify harmful effects of ENPs, research into new methods is required to address the special properties of nanomaterials. It is crucially important to assess their safety for sustainable implementation of nanotechnology with its full potential.

REFERENCES

[1] Ju-Nam Y, Lead JR. Manufactured nanoparticles: an overview of their chemistry, interactions and potential environmental implications. Sci Total Environ 2008; 400: 396-414.

[2] Yang J, Alemany LB, Driver J, Hartgerink JD, Barron AR. Fullerene-derivatized amino acids: synthesis, characterization, antioxidant properties, and solid-phase peptide synthesis. Chemistry 2007; 13: 2530-45.

[3] Aitken RJ, Chaudhry MQ, Boxall AB, Hull M. Manufacture and use of nanomaterials: current status in the UK and global trends. Occup Med (Lond). 2006; 56: 300-6.

[4] Chaudhry Q, Scotter M, Blackburn J, *et al.* Applications and implications of nanotechnologies for the food sector. Food Addit Contam Part A Chem Anal Control Expo Risk Assess 2008; 25: 241-58.

[5] Brigger I, Dubernet C, Couvreur P. Nanoparticles in cancer therapy and diagnosis. Adv Drug Deliv Rev 2002; 54: 631-51.

[6] Panyam J, Labhasetwar V. Biodegradable nanoparticles for drug and gene delivery to cells and tissue. Adv Drug Deliv Rev 2003; 55: 329-47.

[7] Cui Z, Lockman PR, Atwood CS, *et al.* Novel D-penicillamine carrying nanoparticles for metal chelation therapy in Alzheimer's and other CNS diseases. Eur J Pharm Biopharm 2005; 59: 263-72.

[8] Dobson J. Nanoscale biogenic iron oxides and neurodegenerative disease. FEBS Lett 2001; 496: 1-5.

[9] Lecaroz C, Gamazo C, Blanco-Prieto MJ. Nanocarriers with gentamicin to treat intracellular pathogens. J Nanosci Nanotechnol. 2006; 6: 3296-302.

[10] Pertuit D, Moulari B, Betz T, Nadaradjane A, Neumann D, Ismaïli L, Refouvelet B, Pellequer Y, Lamprecht A. 5-amino salicylic acid bound nanoparticles for the therapy of inflammatory bowel disease. J Control Release. 2007; 123: 211-8.

[11] Tsai CY, Shiau AL, Chen SY, *et al.* Amelioration of collagen-induced arthritis in rats by nanogold. Arthritis Rheum. 2007; 56: 544-54.

[12] Hoyt VW, Mason E. Nanotechnology-emerging health issues. J Chem Health Safety 2008; 10: 10-15.

[13]   Casals E, Va´zquez-Campos S, Bastus NG, Puntes V. Distribution and potential toxicity of engineered inorganic nanoparticles and carbon nanostructures in biological systems. Trend Anal Chem 2008; 27: 672-83.

[14]   Kagan VE, Bayir H, Shvedova AA. Nanomedicine and nanotoxicology: two sides of the same coin. Nanomedicine. 2005; 1: 313-6.

[15]   BSI British Standards, PAS 71: Vocabulary - Nanoparticles, 2005.

[16]   Nowack B, Bucheli TD. Occurrence, behavior and effects of nanoparticles in the environment. Environ Pollut. 2007; 150: 5-22.

[17]   Poole CP, Owens FJ. Introduction to Nanotechnology. Hoboken: Wiley-Interscience; 2003.

[18]   Daniel MC, Astruc D. Gold nanoparticles: assembly, supramolecular chemistry, quantum-size-related properties, and applications toward biology, catalysis, and nanotechnology. Chem Rev 2004; 104: 293-346.

[19]   Roco M, Brainbridge WS. Societal Implications of Nanoscience and Nanotechnology. Kluwer Academic Pubishers; 2001.

[20]   Daughton CG. Non-regulated water contaminants: emerging research. Environ Impact Asses Rev 2004; 24: 711-32.

[21]   Howard CV. Small particles-big problems. Int Lab News 2004; 34: 28-9.

[22]   Brook RD. Inhalation of fine particulate air pollution and ozone causes acute arterial vasoconstriction in healthy adults. Circulation 2002; 105: 1534-6.

[23]   Donaldson K, Stone V, Gilmour PS, Brown DM, MacNee W. Ultrafine particles: mechanisms of lung injury. Philos Trans R Soc Lond 2000; 358: 2741-9.

[24]   Lam CW, James JT, McCluskey R, Hunter RL. Pulmonary toxicity of single wall carbon nanotubes in mice 7 and 90 days after intratracheal instillation. Toxicol Sci 2004; 77: 126-34.

[25]   Oberdörster G, Sharp Z, Atudorei V, *et al.* Translocation of inhaled ultrafine particles to the brain. Inhal Toxicol 2004;16:437-45.

[26]   Nowack B, Bucheli TD. Occurrence, behavior and effects of nanoparticles in the environment. Environ Pollut. 2007; 150: 5-22.

[27]   Warheit DB, Sayes CM, Reed KL, Swain KA. Health effects related to nanoparticle exposures: environmental, health and safety considerations for assessing hazards and risks. Pharmacol Ther 2008; 120: 35-42.

[28]   Dowling A. Development of nanotechnologies. Mater Today 2004; 7: 30-5.

[29]   Warheit DB. Nanoparticles: health impacts? Mater Today 2004; 7: 32-5.

[30]   Colvin VL. Sustainability for nanotechnology. The Scientist 2004; 18: 26-7.

[31]   NIH Nanotechnology and Nanoscience Information, National Institutes for Health, United States Department of Health and Human Services, 17 November 2006.

[32]   Oberdörster E. Manufactured nanomaterials (fullerenes, C60) induce oxidative stress in the brain of juvenile largemouth bass. Environ Health Perspect 2004; 112: 1058-62.

[33]   Donaldson K, Tran L, Jimenez LA, *et al.* Combustion-derived nanoparticles: a review of their toxicology following inhalation exposure. Part Fibre Toxicol 2005; 2: 10.

[34]   Elder A, Gelein R, Silva V, *et al.* Translocation of inhaled ultrafine manganese oxide particles to the central nervous system. Environ Health Perspect. 2006; 114: 1172-8.

[35]   Na K, Lee TB, Park K-H, Shin E-K, Lee Y-B, Choi H-K. Self-assembled nanoparticles of hydrophobically modified polysaccharide bearing vitamin H as a targeted anti-cancer drug delivery system. Eur J Pharm Sci 2003; 18: 165-73.

[36]   Panyam J, Sahoo SK, Prabha S, Bargar T, Labhasetwar V. Fluorescence and electron microscopy probes for cellular and tissue uptake of poly (D, L-lactideco-glycolide) nanoparticle. Int J Pharm 2003; 262: 1-11.

[37]   Pelkmans L, Helenius A. Endocytosis *via* caveolae. Traffic 2002;3:311-20.

[38]   Berry CC, Wells S, Charles S, Aitchison G, Curtis ASG. Cell response to dextran-derivatised iron oxide nanoparticles post internalization. Biomaterials 2004; 25: 5405-13.

[39]   Rothen-Rutishauser B, Mühlfeld C, Blank F, Musso C, Gehr P. Translocation of particles and inflammatory responses after exposure to fine particles and nanoparticles in an epithelial airway model. Part Fibre Toxicol 2007; 4: 9.

[40]   Warheit DB, Webb TR, Colvin VL, Reed KL, Sayes CM. Pulmonary bioassay studies with nanoscale and fine quartz particles in rats: toxicity is not dependent upon particle size but on surface characteristics. Toxicol Sci 2007; 95: 270-280.

[41]   Gojova A, Guo B, Kota RS, Rutledge JC, Kennedy IM, Barakat AI. Induction of inflammation in vascular endothelial cells by metal oxide nanoparticles: effect of particle composition. Environ Health Perspect 2007; 115: 403-9.

[42]   Wagner AJ, Bleckmann CA, Murdock RC, Schrand AM, Schlager JJ, Hussain SM. Cellular interaction of different forms of aluminum nanoparticles in rat alveolar macrophages. J Phys Chem B 2007; 111: 7353-9.

[43] Hoet PHM, Brüske-Hohlfeld I, Salata OV. Nanoparticles-known and unknown health risks. J Nanobiotechnol 2004; 2: 12.

[44] Brown DM, Wilson MR, MacNee W, Stone V, Donaldson K. Size-dependent proinflammatory effects of ultrafine polystyrene particles: a role for surface area and oxidative stress in the enhanced activity of ultrafines. Toxicol Appl Pharmacol 2001; 175: 191-9.

[45] Moore MN. Do nanoparticles present ecotoxicological risks for the health of the aquatic environment? Environ Int. 2006; 32: 967-76.

[46] Warheit DB, Webb TR, Reed KL, Frerichs S, Sayes CM. Pulmonary toxicity study in rats with three forms of ultrafine-TiO2 particles: differential responses related to surface properties. Toxicology 2007; 230: 90-104.

[47] Semmler M, Seitz J, Erbe F, *et al.* Long-term clearance kinetics of inhaled ultrafine insoluble iridium particles from the rat lung, including transient translocation into secondary organs. Inhal Toxicol 2004; 16: 453-9.

[48] Oberdörster G, Oberdörster E, Oberdörster J. Nanotoxicology: an emerging discipline evolving from studies of ultrafine particles. Environ Health Perspect. 2005;113:823-39.

[49] Rothen-Rutishauser BM, Schürch S, Haenni B, Kapp N, Gehr P. Interaction of fine particles and nanoparticles with red blood cells visualized with advanced microscopic techniques. Environ Sci Technol 2006; 40: 4353-9.

[50] Brown JS, Zeman KL, Bennett WD. Ultrafine particle deposition and clearance in the healthy and obstructed lung. Am J Respir Crit Care Med 2002; 166: 1240-7.

[51] Renwick LC, Brown D, Clouter A, Donaldson K. Increased inflammation and altered macrophage chemotactic responses caused by two ultrafine particle types. Occup Environ Med 2004; 61: 442-7.

[52] Kaminskas LM, Boyd BJ, Karellas P, Henderson SA, Giannis MP, Krippner GY, Porter CJ. Impact of surface derivatization of poly-L-lysine dendrimers with anionic arylsulfonate or succinate groups on intravenous pharmacokinetics and disposition. Mol Pharm 2007; 4: 949-61.

[53] Raloff J. Air sickness: how microscopic dust particles cause subtle but serious harm. Sci News 2003; 164: 1-11.

[54] Lam CW, James JT, McCluskey R, Arepalli S, Hunter RL. A review of carbon nanotube toxicity and assessment of potential occupational and environmental health risks. Crit Rev Toxicol 2006; 36: 189-217.

[55] Donaldson K, Aitken R, Tran L, *et al.* Carbon nanotubes: a review of their properties in relation to pulmonary toxicology and workplace safety. Toxicol Sci 2006; 92: 5-22.

[56] Cheng XK, Kan AT, Tomsom MB. Naphthalene adsorption and desorption from aqueous C-60 fullerene. J Chem Eng Data 2004; 49: 675-83.

[57] Jia G, Wang H, Yan L, *et al.* Cytotoxicity of carbon nanomaterials: single-wall nanotube, multi-wall nanotube, and fullerene. Environ Sci Technol 2005; 39: 1378-83.

[58] Inoue K, Takano H, Yanagisawa R, *et al.* Effects of inhaled nanoparticles on acute lung injury induced by lipopolysaccharide in mice. Toxicology 2007; 238: 99-110.

[59] Reeves JF, Davies SJ, Dodd NJ, Jha AN. Hydroxyl radicals (*OH) are associated with titanium dioxide (TiO$_2$) nanoparticle-induced cytotoxicity and oxidative DNA damage in fish cells. Mutat Res. 2008; 640: 113-22.

[60] Kaewamatawong T, Shimada A, Okajima M, *et al.* Acute and subacute pulmonary toxicity of low dose of ultrafine colloidal silica particles in mice after intratracheal instillation. Toxicol Pathol. 2006; 34: 958-65.

[61] Xia T, Kovochich M, Brant J, *et al.* Comparison of the abilities of ambient and manufactured nanoparticles to induce cellular toxicity according to an oxidative stress paradigm. Nano Lett 2006; 6: 1794-807.

[62] Wilson MR, Foucaud L, Barlow PG, *et al.* Nanoparticle interactions with zinc and iron: implications for toxicology and inflammation. Toxicol Appl Pharmacol 2007; 225: 80-9.

[63] Waldman WJ, Kristovich R, Knight DA, Dutta PK. Inflammatory properties of iron-containing carbon nanoparticles. Chem Res Toxicol 2007; 20: 1149-54.

[64] Sayes CM, Reed KL, Warheit DB. Assessing toxicity of fine and nanoparticles: comparing *in vitro* measurements to *in vivo* pulmonary toxicity profiles. Toxicol Sci 2007;97:163-80.

[65] Daughton C, Ternes T. Pharmaceuticals and personal care products in the environment: agents of subtle change? Environ Health Perspect 1999; 107: 907-938.

[66] Lázaro FJ, Abadía AR, Romero MS, *et al.* Magnetic characterisation of rat muscle tissues after subcutaneous iron dextran injection. Biochim Biophys Acta 2005; 1740: 434-45.

[67] Kirchner C, Liedl T, Kudera S, Pellegrino T, Muñoz Javier A, Gaub HE, Stölzle S, Fertig N, Parak WJ. Cytotoxicity of colloidal CdSe and CdSe/ZnS nanoparticles. Nano Lett 2005; 5: 331-8.

[68] Auger F, Gendron MC, Chamot C, *et al.* Responses of well-differentiated nasal epithelial cells exposed to particles: role of the epithelium in airway inflammation. Toxicol Appl Pharmacol 2006; 215: 285-94.

[69]   Aitken RJ, Creely KS, Tran CL. Nanoparticles: an occupational hygiene review. Institute of Occupational Medicine for the Health and Safety Executive, London: 2004.

[70]   Liu WT. Nanoparticles and their biological and environmental applications. J Biosci Bioeng 2006; 102: 1-7.

[71]   Royal Society and Royal Academy of Engineering. Nanoscience and nanotechnologies: opportunities and uncertainties. RS policy document 19/04. London: The Royal Society; 2004.

[72]   Tiede K, Hassellöv M, Breitbarth E, Chaudhry Q, Boxall AB. Considerations for environmental fate and ecotoxicity testing to support environmental risk assessments for engineered nanoparticles. J Chromatogr A 2009; 1216: 503-9.